# Assessment & Treatment of ADHD

AF865857

Ryan J. Kennedy
Editor

# Assessment & Treatment of ADHD

## A Practical Guide for Nurse Practitioners

Springer

*Editor*
Ryan J. Kennedy
Pediatric & Adult Clinic for ADHD & Related Disorders
Manhattan Beach, CA, USA

ISBN 978-3-032-24170-2 ISBN 978-3-032-24171-9 (eBook)
https://doi.org/10.1007/978-3-032-24171-9

© The Editor(s) (if applicable) and The Author(s), under exclusive license to Springer Nature Switzerland AG 2026

This work is subject to copyright. All rights are solely and exclusively licensed by the Publisher, whether the whole or part of the material is concerned, specifically the rights of translation, reprinting, reuse of illustrations, recitation, broadcasting, reproduction on microfilms or in any other physical way, and transmission or information storage and retrieval, electronic adaptation, computer software, or by similar or dissimilar methodology now known or hereafter developed.
The use of general descriptive names, registered names, trademarks, service marks, etc. in this publication does not imply, even in the absence of a specific statement, that such names are exempt from the relevant protective laws and regulations and therefore free for general use.
The publisher, the authors and the editors are safe to assume that the advice and information in this book are believed to be true and accurate at the date of publication. Neither the publisher nor the authors or the editors give a warranty, expressed or implied, with respect to the material contained herein or for any errors or omissions that may have been made. The publisher remains neutral with regard to jurisdictional claims in published maps and institutional affiliations.

Editorial Contact: Marie-Elia Come-Garry
This Springer imprint is published by the registered company Springer Nature Switzerland AG
The registered company address is: Gewerbestrasse 11, 6330 Cham, Switzerland

If disposing of this product, please recycle the paper.

# Preface

## The Evolving Role of Nurse Practitioners in ADHD in the United States

Nurse practitioners (NPs) have an important role in the diagnosis and treatment of Attention-Deficit/Hyperactivity Disorder (ADHD) across the United States. In 2022, NPs prescribed more stimulant medications than any other clinician group, accounting for 23% of all ADHD prescriptions, surpassing psychiatrists for the first time (IQVIA Government Solutions, Inc. 2023). This shift reflects a rapidly changing healthcare landscape marked by expanded NP practice authority, increased demand for mental health services, and heightened public scrutiny surrounding ADHD diagnosis and treatment practices. The COVID-19 pandemic further accelerated these trends, contributing to a documented rise in adult ADHD diagnoses—especially among women—and a surge in telehealth-based care (Auro et al. 2024). As NP responsibilities continue to grow, so does the need for advanced clinical competence, ethical practice, and adherence to core nursing values of trust, communication, and safety.

This book aims to provide students and clinicians with practical, up-to-date information and evidence-based guidance for assessment and treatment of ADHD. It addresses common gaps in NP training by offering guidance for integrating ADHD into nursing education, policy and law, and managing both routine and complex clinical scenarios. This book draws from established clinical guidelines, expert consensus, and over 15 years of specialized ADHD practice developed during my mentorship and experience over the past 15 years working in research and clinical practice at the Brown Clinic for Attention and Related Disorders. I worked alongside Thomas E. Brown, PhD, a pioneer and world-renowned expert in ADHD.

The authors contributing to this book represent a diverse and distinguished group of leaders and educators from across the United States, including university professors, psychiatric and family nurse practitioners, national ADHD organization leaders, executive-function coaches, therapists, and clinicians practicing in a wide range of healthcare and academic settings. This multidisciplinary authorship brings a broad and informed perspective on the complexities of supporting individuals with ADHD across the lifespan. Together, their collective expertise highlights the many ways nurse practitioners can meaningfully contribute to high-quality ADHD assessment, treatment, and long-term management. The breadth of professional backgrounds represented in this volume ensures a rich, comprehensive understanding of ADHD care, making this book an authoritative resource for students, clinicians, and healthcare leaders seeking to elevate practice.

The content of this textbook also can be used for integrating ADHD education into professional nursing curricula. One way to achieve this is by incorporating quality improvement (QI) principles into courses that support effective ADHD management in nurse practitioner-led settings. For instance, students or clinicians interested in evaluating clinic workflow, staff roles, medication management policies, patient capacity, and follow-up systems can enhance safety, efficiency, and overall care quality. The Agency for Healthcare Research and Quality (AHRQ) offers many accessible resources for improving communication, streamlining processes, and strengthening patient engagement. The appendix of this book provides sample workflows, checklists, titration schedules, and follow-up protocols to support consistency and standardization in clinical practice.

Finally, this book underscores the importance of psychoeducation, collaborative care models, as well as legal principles involved in safe and effective care. Providing ADHD care requires more than accurate prescribing: it depends on ongoing patient education, transparent communication about expectations and side effects, structured follow-up, and reinforcement of behavioral strategies. These elements serve to empower patients, reduce stigma, and improve adherence. Together, these frameworks equip nurse practitioners to deliver high-quality, ethically grounded, evidence-based ADHD care within an evolving and increasingly complex healthcare system.

## How Can Nurse Practitioners Support Patients' ADHD Treatment?

The expansion of NP scope of practice reflects important changes in our healthcare system, as states in the United States are passing legislation to remove practice restrictions for NPs. However, with expanded prescriptive authority comes the responsibility to uphold high standards of care. As the scope of practice for NPs continues to be debated, it is important that our care for patients maintains core nursing values of trust, effective communication, and safety. These values form the foundation for strong relationships with patients, colleagues, and the public perception of NPs. Such relationships enable NPs to pursue leadership roles and positively influence the healthcare system

Opportunities for growth in the NP profession have also led to heightened scrutiny regarding the delivery of ADHD care. During the COVID-19 pandemic, there was an increased rate of adult ADHD diagnoses, including many women who are identified later in life, which is a growing area of research and treatment. The unprecedented surge in telehealth has raised concerns about rising stimulant prescription rates. While many family and pediatric NPs have less formal psychiatry training, they have the necessary qualifications to provide care to ADHD patients, and can expand on their education to improve their knowledge and skills to be part of the solution. Other practitioners may also feel that their professional education and training did not provide adequate understanding for how to assess and treat patients with ADHD due to limited curriculum or clinical supervision in this area, but as the healthcare landscape evolves, it is essential for us to adapt and pursue additional training to effectively support our patients.

Manhattan Beach, CA, USA Ryan J. Kennedy

## References

Auro, K., Holopainen, I., Perola, M., Havulinna, A. S., & Raevuori, A. (2024). Attention-Deficit/Hyperactivity Disorder Diagnoses in Finland During the COVID-19 Pandemic. JAMA network open, 7(6), e2418204. https://doi.org/10.1001/jamanetworkopen.2024.18204.

IQVIA Institute (2023). Stimulant Prescription Trends in the United States (2012–2022). Report prepared for the DEA. (This report provides epidemiologic data on prescribing; relevant stat cited in introduction).

# Contents

# The Importance of Recognizing ADHD 1

Jane E. Indergaard

In advanced nursing and medical training, diagnoses and conditions are emphasized based on prevalence, severity, complexity, and the potential for impact on patient outcomes. Those typically explored with depth in a curriculum are either common or cause significant impairment and mortality if left unmanaged. Examples of such conditions include heart disease, cancer, hypertension, and diabetes mellitus (Boehning and Punsalan 2025; National Institutes of Health (NIH) 2023; Bryan 2024).

Attention deficit hyperactivity disorder (ADHD) is described as the most common pediatric behavioral health diagnosis in the United States (US). ADHD affects over 7 million children ages 3–17 years or 11.7% (Centers for Disease Control (CDC) 2024). Moreover, ADHD often persists in adulthood affecting a reported 15.5 million (6%) or one in 16 adults (Staley et al. 2024). It has an estimated worldwide prevalence of 8% in children and adolescents (Ayano et al. 2023) and 2.8% in adults (O'Nions et al. 2025). ADHD is a recognized neurodevelopmental condition that, while highly variable in severity, can have far-reaching negative outcomes impacting every life domain, such that it is recognized as a public health issue (Barkley 2020). It is a highly comorbid condition and rarely occurs in isolation, meaning it typically presents with multidimensional stressors for patients and their families. With such a high prevalence and comorbid behavioral and physical associations, it is likely that every advanced nursing practice provider will regularly see clients with ADHD.

Despite recent knowledge of its potential for significant impairment, its highly comorbid associations, and developments in effective multi-modal treatments, ADHD still does not receive the degree of focus, training, and attention it deserves. It can be argued that ADHD should be as much of a mainstay in the training and

J. E. Indergaard (✉)
Sanford Heimarck School of Health Professions, Concordia College, Moorhead, MN, USA
e-mail: indergaa@cord.edu

© The Author(s), under exclusive license to Springer Nature Switzerland AG 2026

R. J. Kennedy (ed.), *Assessment & Treatment of ADHD*,
https://doi.org/10.1007/978-3-032-24171-9_1

education of healthcare providers, as does heart disease, hypertension, diabetes, and depression.

We have learned much about ADHD in the last two decades, and new frontiers of knowledge about its impact across the lifespan and among various demographics continue to emerge. As a common, complex, highly variable, and life-long condition, it is essential that providers are armed with the most recent knowledge about the condition, its diagnosis, and evidence-based treatment modalities. This textbook serves to provide the nurse practitioner (NP) with a comprehensive understanding of a highly misunderstood condition, as well as give practical insights that can be applied directly to clinical practice. In this chapter, we will explore three key points about the condition that will lay the foundation for the importance of recognizing ADHD in clinical practice. These are (1) ADHD is real; (2) ADHD is more than a "nuisance" disorder; and (3) The role of NP in ADHD diagnosis, treatment, and management. Equipped with the right knowledge and skills, the advanced practice nurse can make a profound difference in the lives of patients with ADHD and their families.

## 1.1 ADHD Is Real

Historically ADHD has been a highly misunderstood condition, laden with myths and misconceptions. Among them are beliefs that ADHD is a "made-up" condition for behaviors related to poor parenting, too much sugar, laziness, and poor self-discipline. It has been characterized in some social media and lay literature as a modern invented label for people who want to medicate a child's "bad" behavior. However, the facts are that ADHD is a real, biologically based neurodevelopmental condition, which is highly heritable. Furthermore, there is strong evidence that it has been recognized for a long, long time.

### 1.1.1 It' Been Around a Long Time...

The fact is ADHD is not new in terms of being identified as a unique, definable condition, though conceptually its nomenclature, behavioral, and etiological descriptions reflect a changing and constantly evolving understanding of this condition over time. According to recorded historical literature, inattentive symptoms similar to those depicted in modern ADHD diagnostic criteria were first described as early as 1798 by Scottish Physician, Sir Alexander Crichton, as an "incapacity of attending with a necessary degree of constancy to any object" (Lange et al. 2010).

Other accounts in literature and medical reports from the eighteenth and nineteenth centuries in Europe and the United States describe noted populations of children presenting with what was described in varying terms as "abnormal degrees" of hyperexcitability, emotional reactivity, overactivity, behavioral disturbance, uncontrollability, distraction, and more (Lange et al. 2010; Martinez-Badía and Martinez-Raga 2015). Hints that a persistent behavioral "type" with symptoms we now know as ADHD were even described in fictional literature and art of the nineteenth century, though not established as a psychiatric disorder. In 1844, German physician and author Heinrich Hoffman, wrote about two characters known as *Fidgety Phil* and *Johnny Look-in-the-air*, who represented children presenting with excessive motor activity, impulsivity, and distractedness. It is believed these characters evolved from his own clinical observations of child behavior in his practice. It has been suggested that these characterizations indicate recognition of an unspecified but distinct behavioral presentation in subsets of children, which have existed across centuries. Others argue that the behaviors depicted are insufficient to draw that conclusion. Regardless, in hindsight and especially in the context of Dr. Hoffman's clinical practice in populations of children and in the mental health field of that time, the similarities between his characters and today's understanding of ADHD are remarkable.

British pediatrician Sir George Frederic Still's definition of this protype in 1902 as an "abnormal defect of moral control" is considered the scientific starting point for today's modern conceptualization of ADHD (Martinez-Badía and Martinez-Raga 2015). Moral control as defined at that time closely aligns with the concept of innate self-regulation, rather than volitional misbehavior. However, up to that time, evidence of deficits of "moral control" was frequently attributed to children with underlying brain injury or damage that resulted in incapacity for reasoning. Most notable and significant for nudging medicine and science to a more accurate understanding of ADHD were Still's observations that the condition was not always associated with underlying brain damage or intellectual disability. Rather these "morbid manifestations" of behavior were also observed in children with no history of brain injury, underlying morbidity, or intellectual impairment.

Since that time, evolving sociological and medical advances laid the groundwork for an ever-expanding investigation into the underlying etiology of a condition that we now know as ADHD. Significantly, not unlike modern times, in the eighteenth and nineteenth centuries, the explanation for what had been deemed a prototype of an unstable or dysregulated nervous system was attributed to a lack of discipline, poor upbringing, character defects, and/or brain damage caused by external forces such as perinatal injury. This persisted into the early twentieth century with brain damage from external forces becoming the most embraced explanation, until another enlightening discovery was made.

In 1937 psychiatrist Charles Bradley inadvertently unveiled the positive effect of stimulant medication on various behavior disorders (Strohl 2011). Using pneumoencephalography to examine structural brain abnormalities in behaviorally disordered children, he found that his subjects developed severe headaches. He turned to the drug Benzedrine (amphetamine) for treatment and found no effect on headache but a profound positive effect on school performance of these children and their behavior. Intrigued, he continued further study, expanding the use of the medication on hyperactive children, which demonstrated similar behavioral improvement.

Bradley's work did open the door for clinical research into psychopharmacological treatment of behavior disorders in children. However, it was also met with some criticism due to drug side effects, the fact that treatment was not curative, and his own concern that a reliance on medication might take away the human connection of holistic treatment (Lange et al. 2010; Strohl 2011). While the findings of both Still and Bradley were largely ignored in the short-term, they served to further underscore for the first time a suggested biological or brain-based origin for the behaviors seen and deemed disordered. These findings eventually became a driving force for further study and research across the next several decades.

The long-standing historical accounts, reports, and descriptions about ADHD-like symptoms in children across centuries and the uncovering of evidence for a biological basis for the condition decry misinformed claims that ADHD is a "new," made-up condition and that ADHD is not *real*. Not only is it "real" but due twentieth century pioneering advances in the field of neuroscience, we now know more about the neurobiological basis for this condition.

## 1.2 ADHD: A Brain-Based Neurodevelopmental Condition

Advances in neuro-imaging techniques, genetic epidemiology, and psychopharmacological studies on behavior and the brain in the past 25 years have unveiled an explosive amount of information on the biological underpinnings of ADHD. While much remains to be discovered, current findings have demonstrated key structural and functional differences in brains affected by ADHD and a strong genetic component.

### 1.2.1 Neuroimaging Findings

The advent of advanced imaging technology such as functional magnetic resonance imaging (fMRI) and positive emission tomography (PET) has revolutionized our understanding of many neuropsychiatric conditions. For the first time in medical history, the brain can be observed not only for its morphology and structural alterations but also for metabolism and activity in real time. Responses of the brain to mental tasks, emotions, and activity can be visualized and measured. Comparisons of any differences can be made between the brains of individuals suspected of being affected by various conditions and those who are not.

Neuroimaging studies of the brains of individuals with ADHD have found maturational, structural, and functional differences compared to their non-affected counterparts. Affected regions in ADHD have been found to be widespread across the cortex. Specific findings include observations of smaller brains overall and reduced volume and activity in key areas including the prefrontal cortex, cerebellum, hippocampus, amygdala, and subcortical regions (Carucci et al. 2023; Hoogman et al. 2017, 2019; Dehghani Firouzabadi et al. 2022). Abnormal activation is frequently noted in brain areas associated with motor control, response inhibition, and attention during different emotional and cognitive tasks. Hypoactivation is seen in signaling networks associated with executive function (EF), reward systems, and attention.

It is important to clarify that brain size does not correlate with intelligence; hence neither does ADHD. Equally important is the understanding that fMRI and other scanning only provide a macro view of the brain's workings. Furthermore, there is an overlap in some of these findings with other neurodevelopmental conditions such as autism, so they are not necessarily specific to ADHD. The high degree of heterogeneity among the expression and severity of ADHD points to a multifarious pathogenesis. Nonetheless, the structural and functional differences identified in specific brain areas through neuroimaging can be traced to its symptoms, challenges, and functional impairments. A brief summary of common neuroimaging findings for ADHD and their significance with the consideration of these previous caveats is presented in Table 1.1.

### 1.2.2 Genetics

ADHD is highly familial. Studies have shown that an estimated 70–80% of ADHD is inherited (Rietz et al. 2021). In fact, ADHD is nearly as heritable a trait as height. This strong genetic component has more recently been linked to specific genes such as GFOD1, CHD13, and ADGRL3, involved in brain development and dopamine regulation (Thapar 2018; Grimm et al. 2020). Twelve genetic regions associated with ADHD have also been identified which are also associated with certain physical and other behavioral health conditions suggesting a possible shared genetic risk. The identification of these regions also suggests that ADHD is not caused by a single gene or group of genes. To be clear, no single, *specific* genetic variant has been consistently linked to the disorder (Rietz et al. 2021). Rather, ADHD is highly polygenic. It is influenced by variable combinations of many genes that have different effects between families and individuals. Inherited variations of these genes increase the risk of developing ADHD and explain why ADHD runs in families.

While understood as predominantly familial, the identified genetic variants associated with ADHD do not fully explain the heritability of the condition. Rather, while evident that it is strongly influenced by genetics, a large body of recent research has also demonstrated associations between ADHD and a number of environmental influences. To date, they are not considered causative but may be correlative (Thapar et al. 2013). Future research is aimed at detecting environmental influences on gene expression, as well as additional genetic contributions in the development of ADHD.

ADHD is also pleiotropic. While not fully understood, some of these genes, which are linked to brain development and functions, such as dopamine regulation appear to overlap with those associated with other psychiatric conditions including autism, bipolar disorder, and schizophrenia. This is not surprising as genetic overlap is extensive among neuropsychiatric disorders (Grimm et al. 2020). The shared genetic risk with other disorders helps explain the high rate of comorbidity of ADHD and other behavioral health conditions.

Though less studied, there is some emerging research demonstrating a similar degree of genetic overlap (or shared genetic factors) of ADHD with certain physical

**Table 1.1** ADHD neuroimaging findings

| Brain region | Function | Alteration | Significance |
|---|---|---|---|
| Overall brain volume | | Reduced volume of Grey and white matter | May be associated with reductions in the volumes of specific brain regions and developmental delays in brain maturation seen in ADHD in childhood |
| Prefrontal cortex | High-order cognitive function<br>Executive function | Smaller volume<br>Slower maturation | May interfere with attention, decision-making, impulse control, working memory, and executive function tasks such as organizing, planning and follow-through |
| Cerebellum | Motor control<br>Motor coordination<br>Attention<br>Emotional regulation | Smaller volume<br>Slower maturation | Difficulties with motor coordination and regulation of attention and emotional processing—may contribute to hyperactivity and inattention |
| Hippocampus | Memory formation<br>Spatial navigation<br>Stress and emotional processing | Smaller volume | Challenges with memory consolidation, and ability to form and retain new information may impact remembering instructions, completing tasks, recalling details, and contextualizing experiences |
| Amygdala | Emotional processing<br>Emotional regulation | Smaller volume<br>Slower maturation<br>Altered activity | Manifests as heightened emotional reactivity, emotional dysregulation, and impulsivity |
| Basal ganglia-caudate nucleus | Motor control<br>Regulation of attention<br>Cognitive flexibility | Reduced volume and activity | Impairment with filtering irrelevant stimuli and sustaining attention on a task. Contributes to distractibility, difficulties with EF, and poor impulse control |
| Nucleus accumbens | Part of the brain's reward system<br>Central role in motivation, reward processing, and reinforcement learning | Altered activity | Difficulties with motivation, delayed gratification, and increased tendency to seek immediate rewards over long-term goals |

Sources: Carucci et al. (2023), Hoogman et al. (2019, 2017) and Dehghani Firouzabadi et al. (2022)

conditions, such as obesity, epilepsy, migraines, digestive issues, and immune dysregulation (Rietz et al. 2021). At this time, it cannot be said that they are linked, cause and effect, or even correlated. Rather, until more is known, this finding might highlight the importance for practitioners to consider assessing for co-occurring physical conditions, especially higher-risk ones when seeing a patient with ADHD. Conversely, a practitioner might consider screening for ADHD in individuals with these physical conditions, should a patient come in with day-to-day challenges or sub-clinical features of ADHD.

### 1.2.3 Nongenetic Etiology

As a gross estimate, 15% of ADHD can be attributed to perinatal factors (both prenatal exposures and birth complications) that detrimentally impact the developing brain. Prenatal factors associated with the development of ADHD include maternal drug and alcohol use, smoking, exposure to toxins, and intrapartum infection (Thapar et al. 2013; Chandail et al. 2024; da Silva et al. 2023). Certain exposure risks are quite significant in terms of their association with ADHD. For example, studies have found that prenatal exposure to alcohol accounts for a fivefold increase in the odds of a child having ADHD (Pagnin et al. 2019; Lees et al. 2020). Birth complications linked to the development of ADHD include premature birth, low birth weight, and birth asphyxia (Roigé-Castellví et al. 2021). The remaining 5% of ADHD has been linked to postnatal factors such as traumatic brain injury (Asarnow et al. 2021), encephalitic infections (Chou et al. 2015), lead poisoning (Thapar et al. 2013), and certain pediatric chemotherapies (Fellah et al. 2019).

### 1.2.4 Pharmacological Evidence (Functional Theories)

A strong indicator in the legitimacy of ADHD as a real disorder with a biological basis is the outcome of pharmacological studies on people with ADHD. Not only have multiple studies with both stimulant and non-stimulant medications demonstrated efficacy in resolution of symptoms with pharmacological treatment, but functional brain imaging studies have unveiled the improved arousal and function in previously under-aroused brain regions and circuitry associated with ADHD (da Silva et al. 2023; Faraone 2018; Spencer et al. 2015; Zimmer 2017). These medications target neurotransmitter systems in the brain. Thus, improvement in the core symptoms of ADHD and improved daily function with pharmacological treatment support the underlying physiological explanation that ADHD involves dysregulation of neurotransmitter systems in the brain.

In conclusion, the evidence is that ADHD is a complex, neurodevelopmental disorder. It is associated with specific structural and functional differences in the brains of affected individuals and alterations in neurotransmitter function. It has a strong genetic component and is highly heritable but cannot be fully explained by genes alone with evidence of environmental influences on its expression. It is highly heterogenous in presentation and associated comorbidity but is highly prevalent and can pose a significant impact on day-to-day functioning, well-being, and achievement of life goals. ADHD is not just a nuisance; it is a condition that demands respect, attention, and support.

## 1.3 ADHD Is More Than a "Nuisance" Disorder

Certain aspects of ADHD, such as energy, creativity, humor, passion, hyperfocus, and resilience have been described by some as gifts or strengths in their lived experience. This is especially true when one's social context plays to those strengths, such as an actor with ADHD having humor, spontaneity, and creativity. However, for a significant portion of affected individuals, ADHD is linked to adverse outcomes and pronounced challenges in day-to-day living.

### 1.3.1 Impact on Quality of Life

As early as 2001, renowned expert and prominent researcher Russell Barkley introduced his model of the eight life domains in relation to ADHD. His longitudinal research highlighted areas most impacted by the symptoms of ADHD, where he found significant degrees of impairment across life domains. Though the study sample was limited to clinically referred adolescent boys followed through early adulthood, the dissemination of these findings played a crucial role in the understanding of ADHD as a lifelong condition that required ongoing management (Barkley 2002).

Since that time, there have been a sobering number of studies across a wide variety of sample demographics, which consistently demonstrate that ADHD increases the risk for an adverse health-related, lifestyle, and psychosocial outcomes across virtually every life domain. The totality of findings is beyond the scope of this chapter. However, using Barkley's eight life domains as a template, a generalized summation of results across studies is presented below. Findings described are those ubiquitous among research but certainly do not represent the full scope of impairments and challenges found in each domain:

1. **Education**—*People with ADHD are at risk to be undereducated according to potential and experience more grade retention and failure in younger grades than the general population. The drop out and incompletion rates for high school and postsecondary education are alarming when compared to non-affected peers; importantly educational challenges limit future earning potential* (Barkley 2002; French et al. 2024; Harstad et al. 2020; Krauss and Schellenberg 2022; Quintero et al. 2019).
2. **Employment**—*People with ADHD experience more frequent unemployment, have more difficulty securing employment, and demonstrate an increased incidence of job changes both from quitting (bored) and being fired (performance) than their non-affected counterparts* (Barkley 2002; French et al. 2024; Harstad et al. 2020; Quintero et al. 2019; Rattay and Robinson 2024). *Notably mothers of children with ADHD (whether having ADHD themselves or not) frequently leave the workforce to care for their affected children* (Callander et al. 2016; Babinski et al. 2020)*; occupational challenges directly impact socioeconomic status and financial security.*

3. **Motor Vehicle Risks**—*There is a significant increase of serious crashes and other adverse driving outcomes, such as speeding tickets and a higher incidence of license suspension associated with ADHD than found in the general public* (Barkley 2002; French et al. 2024; Harstad et al. 2020).
4. **Financial Management**—*Individuals with ADHD overall are more financially dependent on others and have lower incomes. Studies on financial management and ADHD reported problems with impulse buying, lower-saving, poor financial knowledge, and judgment* (Koerts et al. 2021).
5. **Social and Lifestyle**—(***Social***) *As a group, children and adults with ADHD experience a higher incidence of relationship issues including more serial relationships, increased relational conflict, peer rejection, higher rates of divorce, higher incidence of interpersonal hostility, and intimate partner violence* (Barkley 2002; French et al. 2024; Quintero et al. 2019). *Having ADHD is associated with earlier offending and an increase in childhood, adolescent, and adult arrests and increased conviction and incarcerations* (Barkley 2002; French et al. 2024; Harstad et al. 2020).

   (***Lifestyle***) *Individuals with ADHD frequently demonstrate a more sedentary lifestyle and poorer health maintenance activities such as lower engagement in exercise, attention to diet, and preventative health practices including dental hygiene. Studies also report a higher incidence of adverse health activities such as smoking and substance use among the ADHD population* (Barkley 2002; French et al. 2024; Quintero et al. 2019).
6. **Sexual Activity**—*People with ADHD have an earlier onset of sexual activity than their non-affected peers, more sexual partners, higher-risk behaviors, increased unplanned pregnancy, and a higher incidence of sexually transmitted infections* (Barkley 2002; French et al. 2024; Harstad et al. 2020).
7. **Health Risks**—***(Physical)*** *The ADHD population demonstrates an increased risk for a host of adverse medical conditions which include obesity (up to 41%), coronary heart disease, sleep problems (up to 64%), traumatic brain injury, poor dental hygiene, type II diabetes, allergies, asthma, migraines, eating pathology, severe premenstrual syndrome, and skin conditions such as eczema* (Barkley 2002; French et al. 2024; Krauss and Schellenberg 2022; Rattay and Robinson 2024). *Additionally, there is recent literature which suggests women with ADHD are at increased risk for a host of perinatal complications* (Baker et al. 2022).

   ***(Psychological)*** *The ADHD population has an increased risk for mood disorders (anxiety, depression), accidental and self-inflicted injuries in childhood and adulthood, increased suicide risk* (French et al. 2024; Harstad et al. 2020; Krauss and Schellenberg 2022; Quintero et al. 2019), *and, in women, an increased incidence of premenstrual dysphoric disorder* (Dorani et al. 2021) *and postpartum depression* (Andersson et al. 2023).
8. ***Substance Use and Abuse***—*Individuals with ADHD are reported to have increased rates of tobacco, alcohol, and marijuana use and higher rates for substance use disorder treatment* (French et al. 2024; Harstad et al. 2020). *Note: The incidence is higher in untreated individuals; recent studies indicate that treatment with ADHD medications appears to be a protective factor against substance use disorder* (McCabe et al. 2023; Srichawla et al. 2022).

As alarming as these findings are, it is important to recognize that not everyone with ADHD will have all or even some of these issues. However, an established body of research across the ADHD demographic clearly establishes that children and adults with ADHD are at increased risk for a range of challenges and impairments, including physical and behavioral health conditions. These can significantly impact the quality of life of affected individuals and their families in varying degrees. At the same time, it is not only the impact of ADHD on quality-of-life that healthcare providers need to understand. While suggestive evidence has existed for approximately two decades, more recent epidemiological research and meta-analyses have reported that not only is there an impact on quality of life, but ADHD is also associated with a shortened lifespan (O'Nions et al. 2025; Barkley 2020, 2022).

### 1.3.2 Life Expectancy

Findings from multiple epidemiological studies and recent meta-analyses suggest that ADHD is associated with premature mortality in both children and adults (O'Nions et al. 2025; Barkley and Dawson 2022). Statistical findings of mortality rates between individuals with ADHD in comparison to the general public include a greater than twofold risk for early mortality in childhood, a fourfold increased risk for mortality by age 45 in adulthood (Catalá-López et al. 2022) and a two-times higher risk for early mortality for adults with ADHD (Cattoi et al. 2021). A 2025 study from the United Kingdom demonstrated a 6.78-year reduction in life expectancy for males with ADHD and an 8.64-year reduction for females when compared to the general population (O'Nions et al. 2025).

To date, studies on premature mortality for children and young adults with ADHD attribute the increased incidence predominantly to preventable unnatural causes (accidental injury, suicide, homicide, motor vehicle crashes) (Barkley and Dawson 2022; Catalá-López et al. 2022). However, the cumulative effect of unhealthy lifestyle practices as well as the presence of multiple adverse physical and behavioral health comorbidities associated with ADHD, though less studied, certainly contributes to early mortality.

Most notably both the natural and unnatural causes that align with ADHD mortality stem from the impact of ADHD's core symptoms on multiple domains over the lifespan. They are malleable, modifiable risk factors. As a result, both quality of life and life expectancy in ADHD could be improved by systematic screening, early detection, and preventative health interventions (Callander et al. 2016). Therefore, it is crucial that ADHD is recognized not just as a behavioral health condition but a public health concern.

### 1.3.3 A Public Health Concern

The magnitude of untreated and unmanaged ADHD as a public health issue cannot be overstated. The significant toll on the well-being of individuals and their families notwithstanding, the economic cost and societal burden of ADHD is far-reaching and not to be ignored (Cattoi et al. 2021). Direct and indirect costs for undertreated and untreated ADHD, like other chronic illnesses, are significantly high. According to Sciberras et al., the estimated economic and social costs of ADHD in 2018 to 2019 in the United States were $12.76 billion or a $15,664 per person cost per year (Sciberras et al. 2022). These costs can be broken down into three areas of economic burden: *Productivity*, *Dead Weight Losses*, and *Health System Costs*.

Productivity costs associated with ADHD stem from lost work output due to issues such as low educational attainment, higher unemployment, disproportionally increased occupational challenges, early mortality, and increased suicidality. These costs represent 81% of the total financial costs of ADHD (Sciberras et al. 2022).

Deadweight losses in ADHD stem from any economic impact beyond direct healthcare costs and can also be associated with early mortality and lost productivity, as well as factors such as reduced tax revenue and the day-to-day nonmedical expenses of meeting the needs for individuals with ADHD. These items include special education, childcare costs, parents leaving the workforce to care for their child, transportation for special services, etc. These costs are estimated to make up 11% of the cost burden in ADHD (Sciberras et al. 2022).

Finally, health system costs, which are much higher than the general population (Witrick et al. 2023), represent 4 % of the total economic burden of ADHD and include provider visits, medications, and other treatments or therapies (Sciberras et al. 2022). Clearly, the risks associated with ADHD across life domains have serious implications, not just for individuals and families but for society at large. As with other chronic illnesses, the demand for recognition and targeted attention is warranted. Because many of the challenging outcomes and impact can be mitigated by early detection and effective treatment, public health practices such as integrated care, systematic screening, and preventative interventions are crucial.

We must invest and advocate for improved identification and treatment of ADHD. We must help lawmakers and policymakers understand it is not just a behavioral health condition but an established and serious public health problem (Cattoi et al. 2021). Lastly, ADHD needs to be a mainstay in all healthcare education and training due to its significant risks for adverse mental and physical health problems, early mortality, and costly social and lifestyle consequences if not effectively managed.

## 1.4 The Role of the NP in ADHD Diagnosis, Treatment, and Management

In the past two decades, new medications, therapies, and treatment approaches have emerged, in the field of ADHD, offering more effective and targeted interventions. Our knowledge about ADHD and tools for screening and diagnosis continue to improve and evolve. However, there is unmistakable evidence that across the lifespan perspective, ADHD often goes undiagnosed and/or untreated. This is particularly true across various demographics which represent significant disparities in care. For example, according to recent CDC statistics, approximately 11.4% of children ages 3–17 have been diagnosed with ADHD in the United States. However, nearly 30% of them reportedly do not receive adequate treatment including medications and behavioral treatments (Danielson et al. 2024). Black and Hispanic children are diagnosed less frequently than their white counterparts, even when socioeconomic status is controlled (Rattay and Robinson 2024). It is estimated that 4.4% of US adults have ADHD, but less than one-third have been diagnosed and 25% of those diagnosed do not receive treatment (Staley et al. 2024). Girls and women with ADHD are diagnosed less often, referred for screening less often and are diagnosed later than their male counterparts (Young et al. 2020).

As with other chronic conditions, contributing factors to these disparities include lack of access and resources associated with a shortage of mental health professionals, poor insurance coverage, geographical barriers, and prohibitive costs of care. But there are added dimensions to disparity in ADHD identification and treatment which include continued stigma and limited education for ADHD as a complex but treatable neurodevelopmental condition.

Understanding ADHD as a lifelong condition requiring ongoing management is a critical perspective for practitioners and providers in all healthcare fields. Whether working in behavioral health, pediatrics, women's health or adult care, ADHD is a common, prevalent condition that intersects with and impacts other conditions. Not only is early detection and intervention important, but awareness of its fluctuating nature across the lifespan and its potential contribution to risks in other health dimensions is essential. Like other pediatric chronic conditions such as diabetes, asthma, or autism, proper care for ADHD requires long-term follow-up, regular review, continuity of care, as well as addressing the need for family education and ongoing support (Kleve et al. 2022).

Nurse practitioners (NP) by virtue of their education, training, scope of practice, and holistic approach to patient care are perfectly positioned to fill in the gaps for equity and quality access for individuals with ADHD. Working in a wide range of practice settings, their footprint and reach among patient populations is large. With specialized knowledge and advanced clinical skills, the NP can provide a direct influence on care and services in every aspect of ADHD management including assessment, diagnostics, medical and nonmedical treatments, psychosocial support, patient, and family education.

## 1.5 Conclusion

ADHD is a complex neurodevelopmental condition with a significant risk for long-term morbidity and mortality. It is among the most common pediatric chronic conditions, is highly comorbid, and persists in adulthood with a fluctuating course across the lifespan. It is common and costly in terms of personal and societal burden. It is doubtful that any healthcare provider will have a practice that does not encounter ADHD. Yet many remain woefully undereducated in terms of this common and significant condition. There are effective treatments and management approaches but access to care remains a public health issue for many subpopulations.

In this chapter, we have built a compelling case for making ADHD a cornerstone of education in the training of healthcare professionals, with a sharp focus on the unique and vital role of the nurse practitioner (NP). The NP is uniquely positioned to offer holistic, comprehensive care and bridge the gap in treatment access for individuals with ADHD and their families. However, to truly make an impact, they must be equipped with the latest, evidence-based knowledge. This guide serves as an essential resource, offering expert-level insights that empower nurse practitioners to confidently recognize, understand, and make informed decisions regarding the care and management of ADHD patients—no matter the practice setting. With this knowledge, NPs will be ready to take on one of the most pressing healthcare challenges of our time.

## References

Andersson A, Garcia-Argibay M, Viktorin A, Ghirardi L, Butwicka A, Skoglund C, et al. Depression and anxiety disorders during the postpartum period in women diagnosed with attention deficit hyperactivity disorder. J Affect Disord. 2023;15(325):817–23. https://doi.org/10.1016/j.jad.2023.01.069. Available from: https://www.sciencedirect.com/science/article/pii/S016503272300085X?via%3Dihub

Asarnow RF, Newman N, Weiss RE, Su E. Association of attention-deficit/hyperactivity disorder diagnoses with pediatric traumatic brain injury: a meta-analysis. JAMA Pediatr. 2021;175(10):1009–16. https://doi.org/10.1001/jamapediatrics.2021.2033. Available at: https://jamanetwork.com/journals/jamapediatrics/fullarticle/2782026

Ayano G, Demelash S, Gizachew Y, Tsegay L, Alati R. The global prevalence of attention deficit hyperactivity disorder in children and adolescents: An umbrella review of meta-analyses. J Affect Disord. 2023;339:860–6. https://doi.org/10.1016/j.jad.2023.07.071. Available from: https://pubmed.ncbi.nlm.nih.gov/37495084/

Babinski DE, Mazzant JR, Merrill BM, Waschbusch DA, Sibley MH, Gnagy EM, et al. Lifetime caregiver strain among mothers of adolescents and young adults with attention-deficit/hyperactivity disorder. J Fam Psychol. 2020;34(3):342–52. https://doi.org/10.1037/fam0000609. Available from: https://pubmed.ncbi.nlm.nih.gov/31750692/

Baker A, Wales R, Noe O, Gaccione P, Freeman M, Cohen L. The course of ADHD during pregnancy. J Atten Disord. 2022;26(2):143–8. https://doi.org/10.1177/1087054720975864. Available from: https://pubmed.ncbi.nlm.nih.gov/33307923/

Barkley RA. Major life activity and health outcomes associated with attention-deficit/hyperactivity disorder. J Clin Psychiatry. 2002;63(Suppl 12):10–5. Available from: https://pubmed.ncbi.nlm.nih.gov/12562056/

Barkley R. ADHD adversely impacts health, mortality risk, and estimated life expectancy by adulthood. ADHD Rep. 2020;28(4). https://doi.org/10.1521/adhd.2020.28.4.1. Available from: https://guilfordjournals.com/doi/pdf/10.1521/adhd.2020.28.4.1

Barkley R. High time preference is a key cognitive deficit in ADHD: impact on daily life, impairments and life expectancy. ADHD Rep. 2022;30(7). https://doi.org/10.1521/adhd.2022.30.7.1. Available from: https://guilfordjournals.com/doi/epdf/10.1521/adhd.2022.30.7.1

Barkley RA, Dawson G. Higher risk of mortality for individuals diagnosed with autism spectrum disorder or attention-deficit/hyperactivity disorder demands a public health prevention strategy. JAMA Pediatr. 2022;176(4):e216398. https://doi.org/10.1001/jamapediatrics.2021.6398. Available from: https://pubmed.ncbi.nlm.nih.gov/35157011/

Boehning AP, Punsalan LD. Advanced practice registered nurse roles. In: StatPearls [Internet]. National Library of Medicine. Treasure Island: StatPearls Publishing; 2025. [Updated 2023 Mar 1]. Available from: https://www.ncbi.nlm.nih.gov/books/NBK589698/.

Bryan A. 10 most common health conditions in the U.S. in 2024 [Internet]. 2024 [cited Jan 1]. Available from: https://www.talktomira.com/post/10-most-common-health-conditions-in-the-u-s

Callander EJ, Allele F, Roberts H, Guinea W, Lindsay DB. The effect of childhood ADD/ADHD on parental workforce participation. J Atten Disord. 2016;23(5):487–92. https://doi.org/10.1177/1087054716680076. Available from: https://journals.sagepub.com/doi/full/10.1177/1087054716680076

Carucci S, Narducci C, Bazzoni M, Balia C, Donno F, Gagliano A, Zuddas A. Clinical characteristics, neuroimaging findings, and neuropsychological functioning in attention-deficit hyperactivity disorder: sex differences. J Neurosci Res. 2023;101(5):704–17. https://onlinelibrary.wiley.com/doi/10.1002/jnr.25038

Catalá-López F, Hutton B, Page MJ, Driver JA, Ridao M, Alonso-Arroyo A, et al. Mortality in persons with autism spectrum disorder or attention-deficit/hyperactivity disorder: a systematic review and meta-analysis. JAMA Pediatr. 2022;176(4):e216401. https://doi.org/10.1001/jamapediatrics.2021.6401. Available from: https://pubmed.ncbi.nlm.nih.gov/35157020/

Cattoi B, Alpern I, Katz JS, Keepnews D, Solanto MV. The adverse health outcomes, economic burden, and public health implications of unmanaged attention deficit hyperactivity disorder (ADHD): a call to action resulting from CHADD summit, Washington, DC, October 17, 2019. J Atten Disord. 2021;26(6):807–8. https://doi.org/10.1177/10870547211036754. Available from: https://journals.sagepub.com/doi/10.1177/10870547211036754. (Original work published 2022)

Centers for Disease Control (CDC). Data and statistics on ADHD. [Internet]. 2024 [cited 2024 Nov 19]. Available from: https://www.cdc.gov/adhd/data/index.html

Chandail A, Sangeeta K, Kuman A. Prenatal exposure and ADHD: assessing the impact of maternal stress, nutrition, and substance use. Int J Neurol Nurs. 2024;10(2):1–8. https://doi.org/10.37628/IJNeN. Available at: https://preventionconversation.org/wp-content/uploads/2025/01/1-8-Final_Prenatal-Exposure-and-ADHD.pdf

Chou IC, Lin CC, Kao CH. Enterovirus encephalitis increases the risk of attention deficit hyperactivity disorder: a Taiwanese population-based case-control study. Medicine (Baltimore). 2015;94(16):e707. https://doi.org/10.1097/MD.0000000000000707. Available from: https://pubmed.ncbi.nlm.nih.gov/25906098/

da Silva BS, Grevet EH, Silva, Ramos JK, Rovaris DL, Dotto Bau CH. An overview on neurobiology and therapeutics of attention-deficit/hyperactivity disorder. Discov Ment Health. 2023;3(2). https://doi.org/10.1007/s44192-022-00030-1. Available at: https://link.springer.com/article/10.1007/s44192-022-00030-1

Danielson ML, Claussen AH, Bitsko RH, Katz SM, Newsome K, Blumberg SJ, et al. ADHD prevalence among U.S. children and adolescents in 2022: diagnosis, severity, co-occurring disorders, and treatment. J Clin Child Adolesc Psychol. 2024;53(3):343–60. https://doi.org/10.1080/15374416.2024.2335625. Available from: https://pubmed.ncbi.nlm.nih.gov/38778436/

Dehghani Firouzabadi F, Ramezanpour S, Dehghani Firouzabadi M, Yousem IJ, Puts N, Yousem MD. Neuroimaging in attention-deficit/hyperactivity disorder: recent advances. AJR. 2022;218. Available at: https://www.ajronline.org/doi/pdfplus/10.2214/AJR.21.26316

Dorani F, Bijlenga D, Beekman ATF, van Someren EJW, Kooij JJS. Prevalence of hormone-related mood disorder symptoms in women with ADHD. J Psychiatr Res. 2021;133:10–5. https://doi.org/10.1016/j.jpsychires.2020.12.005. Available from: https://pubmed.ncbi.nlm.nih.gov/33302160/

Faraone SV. The pharmacology of amphetamine and methylphenidate: relevance to the neurobiology of attention-deficit/hyperactivity disorder and other psychiatric comorbidities. Neurosci Biobehav Rev. 2018;87:255–70. https://doi.org/10.1016/j.neubiorev.2018.02.001.

Fellah S, Cheung YT, Scoggins MA, Zou P, Sabin N, Pui CH, et al. Brain activity associated with attention deficits following chemotherapy for childhood acute lymphoblastic leukemia. JNCI. 2019;111(2):201–9. Available from: https://doi.org/10.1093/jnci/djy089.

French B, Nalbant G, Wright H, Sayal K, Daley D, Groom M, et al. The impacts associated with having ADHD: an umbrella review. Front Psychiatry. 2024;15:1343314. https://doi.org/10.3389/fpsyt.2024.1343314.

Grimm O, Kranz TM, Reif A. Genetics of ADHD: what should the clinician know? Curr Psychiatry Rep. 2020;22:18. https://doi.org/10.1007/s11920-020-1141-x. Available at: https://link.springer.com/article/10.1007/s11920-020-1141-x#citeas

Harstad EB, Katusic S, Sideridis G, Weave AL, Voigt RG, Barbaresi WJ. Children with ADHD are at risk for a broad array of adverse adult outcomes that cross functional domains: results from a population-based birth cohort study. J Atten Disord. 2020;26(1):3–14. https://doi.org/10.1177/1087054720964578.

Hoogman M, Bralten J, Hibar DP, Mennes M, Zwiers MP, Schweren LSJ, et al. Subcortical brain volume differences in participants with attention deficit hyperactivity disorder in children and adults: a cross-sectional mega-analysis. Lancet Psychiatry. 2017;4(4):310–9. https://doi.org/10.1016/S2215-0366(17)30049-4. Available from: https://pubmed.ncbi.nlm.nih.gov/28219628/

Hoogman M, Muetzel R, Guimaraes JP, Shumskaya E, Mennes M, Zwiers MP, et al. Brain imaging of the cortex in ADHD: a coordinated analysis of large-scale clinical and population-based samples. Am J Psychiatry. 2019;176(7):531–42. https://doi.org/10.1176/appi.ajp.2019.18091033. Available from: https://psychiatryonline.org/doi/10.1176/appi.ajp.2019.18091033

Kleve L, Vårdal L, Elgen IB. The nurse role in the management of ADHD in children and adolescent: a literature review. Front Psychiatry. 2022;13:676528. https://doi.org/10.3389/fpsyt.2022.676528. Available from: https://www.frontiersin.org/journals/psychiatry/articles/10.3389/fpsyt.2022.676528/full. PMID: 35273525; PMCID: PMC8902137.

Koerts J, Bangma DF, Fuermaier AB, Mette C, Tucha L, Tucha O. Financial judgment determination in adults with ADHD. J Neural Transm. 2021;128:969–79. https://doi.org/10.1007/s00702-021-02323-1.

Krauss A, Schellenberg C. ADHD symptoms and health-related quality of life of adolescents and young adults. Eur J Health Psychol. 2022;29(4):165–74. https://doi.org/10.1027/2512-8442/a000104. Available from: https://econtent.hogrefe.com/doi/10.1027/2512-8442/a000104

Lange KW, Reichl S, Lange KM, et al. The history of attention deficit hyperactivity disorder. ADHD Atten Def Hyp Disord. 2010;2:241–55. https://doi.org/10.1007/s12402-010-0045-8.

Lees B, Mewton L, Jacobus J, Valadez EA, Stapinski LA, Teesson M, et al. Association of prenatal alcohol exposure with psychological, behavioral, and neurodevelopmental outcomes in children from the adolescent brain cognitive development study. Am J Psychiatry. 2020;177(11):1060–72. https://doi.org/10.1176/appi.ajp.2020.20010086. Epub 2020 Sep 25. PMID: 32972200; PMCID: PMC7924902.

Martinez-Badía J, Martinez-Raga J. Who says this is a modern disorder? The early history of attention deficit hyperactivity disorder. World J Psychiatry. 2015;5(4):379–86. https://doi.org/10.5498/wjp.v5.i4.379. Available from: https://pubmed.ncbi.nlm.nih.gov/26740929/

McCabe SE, Schulenberg JE, Wilens TE, Schepis TS, McCabe VV, Veliz PT. Cocaine or methamphetamine use during young adulthood following stimulant use for attention-deficit/hyperactivity disorder during adolescence. JAMA Netw Open. 2023;6(7):e2322650. https://doi.org/10.1001/jamanetworkopen.2023.22650. Available from: https://jamanetwork.com/journals/jamanetworkopen/fullarticle/2807139

National Institutes of Health (NIH). Our biggest health challenges. [Internet]. 2023. Available from: https://www.nih.gov/about-nih/what-we-do/nih-turning-discovery-into-health/our-biggest-health-challenges#:~:text=Heart%20disease%20and%20stroke%20still,understand%20and%20manage%20these%20disorders

O'Nions E, El Baou C, John A, Lewer D, Mandy W, McKechnie DGJ, et al. Life expectancy and years of life lost for adults with diagnosed ADHD in the UK: matched cohort study. Br J Psychiatry. 2025:1–8. https://doi.org/10.1192/bjp.2024.199. Available from: https://pubmed.ncbi.nlm.nih.gov/39844532/

Pagnin D, Zamboni Grecco ML, Furtado EF. Prenatal alcohol use as a risk for attention-deficit/hyperactivity disorder. Eur Arch Psychiatry Clin Neurosci. 2019;269(6):681–7. https://doi.org/10.1007/s00406-018-0946-7. Epub 2018 Oct 23.

Quintero J, Morales I, Vera R, Zuluaga P, Fernández A. The impact of adult ADHD in the quality of life profile. J Atten Disord. 2019;23(9):1007–16. https://doi.org/10.1177/1087054717733046. Available from: https://pubmed.ncbi.nlm.nih.gov/28974134/

Rattay K, Robinson LR. Identifying risk factors for attention-deficit/hyperactivity disorder (ADHD): a public health concern and opportunity. Prev Sci. 2024;25(Suppl 2):195–202. https://doi.org/10.1007/s11121-024-01667-w. Available from: https://pubmed.ncbi.nlm.nih.gov/38598041/

Rietz E, Brikell I, Agnieszka B, Leone M, Chang Z, Cortese S, et al. Mapping phenotypic and aetiological associations between ADHD and physical conditions in adulthood in Sweden: a geneticallyl informed register study. Lancet Psychiatry. 2021;8(9):774–83. https://doi.org/10.1016/S2215-0366(21)00171-1. Available at: https://www.thelancet.com/journals/lanpsy/article/PIIS2215-0366(21)00171-1/fulltext

Roigé-Castellví J, Morales-Hidalgo P, Voltas N, Hernandez-Martinez C, van Ginkel G, Canals J. Prenatal and perinatal factors associated with ADHD risk in schoolchildren: EPINED epidemiological study. Eur Child Adolesc Psychiatry. 2021;30:347–58. https://doi.org/10.1007/s00787-020-01519-2. Available at: https://link.springer.com/article/10.1007/s00787-020-01519-2#citeas

Sciberras E, Streatfeild J, Ceccato T, Pezzullo L, Scott JG, Middeldorp CM, et al. Social and economic costs of attention-deficit/hyperactivity disorder across the lifespan. J Atten Disord. 2022;26(1):72–87. https://doi.org/10.1177/1087054720961828. Available from: https://pubmed.ncbi.nlm.nih.gov/33047627/

Spencer RC, Devilbiss DM, Berridge CW. The cognition-enhancing effects of psychostimulants involve direct action in the prefrontal cortex. Biol Psychiatry. 2015;77:940–50. https://doi.org/10.1016/j.biopsych.2014.09.013.

Srichawla BS, Telles CC, Schweitzer M, Darwish B. Attention deficit hyperactivity disorder and substance use disorder: a narrative review. Cureus. 2022;14(4):e24068. https://doi.org/10.7759/cureus.24068. Available from: https://pubmed.ncbi.nlm.nih.gov/35573587/

Staley BS, Robinson LR, Claussen AH, Katz SM, Danielson ML, Summers AD, et al. Attention-deficit/hyperactivity disorder diagnosis, treatment, and telehealth use in adults. MMWR Morb Mortal Wkly Rep. 2024;73(40). Available from: https://www.cdc.gov/mmwr/volumes/73/wr/pdfs/mm7340-H.pdf

Strohl M. Bradley's Benzedrine studies on children with behavioral disorders. Yale J Biol Med. 2011;84(1):27–31. Available from: https://pmc.ncbi.nlm.nih.gov/articles/PMC3064242/pdf/yjbm_84_1_27.pdf

Thapar A Discoveries on the genetics of ADHD in the 21st century: new findings and their implications. Am J Psychiatry 2018; 175(10). https://doi.org/10.1176/appi.ajp.2018.18040383. Available at: https://psychiatryonline.org/doi/10.1176/appi.ajp.2018.18040383

Thapar A, Cooper M, Eyre O, Langley K. Practitioner review: what have we learnt about the causes of ADHD? J Child Psychol Psychiatry. 2013;54:3–16. https://doi.org/10.1111/j.1469-7610.2012.02611.x. Available at: https://acamh.onlinelibrary.wiley.com/doi/10.1111/j.1469-7610.2012.02611.x

Witrick B, Zhang D, Su D, Li Y, McCall WV, Hendricks B, et al. Medical expenditures associated with attention-deficit/hyperactivity disorder among adults in the United States by age, 2015–2019. J Gen Intern Med. 2023;38(9):2082–90. https://doi.org/10.1007/s11606-023-08075-w. Available from: https://pubmed.ncbi.nlm.nih.gov/36781580/

Young S, Adamo N, Ásgeirsdóttir BB, Branney P, Beckett M, Colley W, et al. Females with ADHD: an expert consensus statement taking a lifespan approach providing guidance for the identification and treatment of attention-deficit/hyperactivity disorder in girls and women. BMC Psychiatry. 2020;20:404. https://doi.org/10.1186/s12888-020-02707-9. Available from: https://bmcpsychiatry.biomedcentral.com/articles/10.1186/s12888-020-02707-9#citeas

Zimmer L. Contribution of clinical neuroimaging to the understanding of the pharmacology of methylphenidate. Trends Pharmacol Sci. 2017;38:608–20. https://doi.org/10.1016/j.tips.2017.04.001.

# 2 ADHD and Mental Health Education for Nurse Practitioners: Strategies to Advance Equity

Roberta Waite and Karan Kverno

## 2.1 Introduction

Attention-deficit/hyperactivity disorder (ADHD) is a neurodevelopmental disorder associated with functional impairment across multiple settings. The prevalence of diagnosed ADHD among children, aged 3–17, based on data collected in the 2022 National Survey of Children's Health, was 10.5%, and of those diagnosed, 77.9% had one or more co-occurring mental, behavioral, or developmental disorder (Danielson et al. 2024). The prevalence of adults, 18 and older, was estimated to be 6% in 2023 based on data collected by the National Center for Health Statistics Rapid Surveys System with 55.9% of reporting that they received their diagnosis in childhood (Staley et al. 2024). The most common comorbidities reported among children were neurodevelopmental, behavioral/conduct, and anxiety disorders (Danielson et al. 2024). If ADHD goes untreated, the risk for comorbidities increases with age, including risks for depression, substance misuse and substance use disorders, accidental injuries, sexually transmitted infections, teenage pregnancy, criminal activity, and mortality (Faraone et al. 2021). Appropriate treatment matters across the lifespan and reduces risks for comorbidities and all-cause mortality, especially mortality from unnatural causes including suicide and accidental injuries (Li et al. 2024).

Most nurse practitioners (NPs) have occasion to evaluate or treat persons with ADHD, especially NPs working in mental health care and primary care settings. Although ADHD is considered a mental disorder, the American Academy of Child and Adolescent Psychiatry estimates that approximately 80% of mental health needs for youth can be effectively addressed through school mental health and primary mental health collaborations in primary care settings (AACAP 2024).

R. Waite (✉) · K. Kverno
Georgetown University, Berkley School of Nursing, Washington, DC, USA
e-mail: rw996@georgetown.edu

© The Author(s), under exclusive license to Springer Nature Switzerland AG 2026
R. J. Kennedy (ed.), *Assessment & Treatment of ADHD*,
https://doi.org/10.1007/978-3-032-24171-9_2

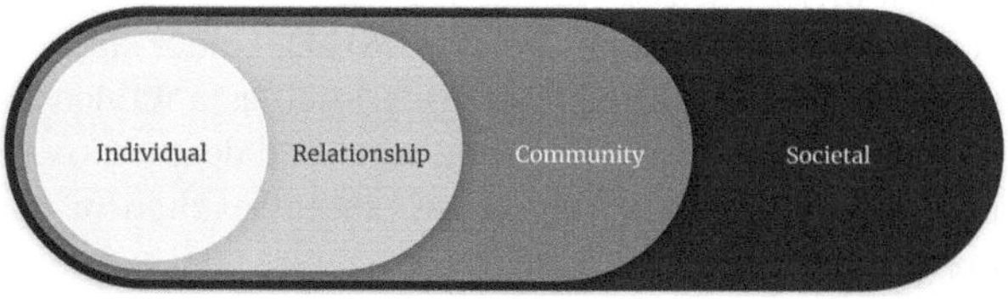


**Fig. 2.1** Social-ecological model. (CDC, public domain)

Traditional NP education requires all students to take physical health assessment, but not all NP education includes sufficient education on evaluation and management of mental health conditions. When nursing schools have a dedicated psychiatric mental health nurse practitioner program, some concepts and content can transcend across NP educational pathways to strengthen mental health competencies in other NP programs. Caring for persons with ADHD requires a strong scientific understanding of genetic, neurobiological, and environmental risk factors, comorbidities, best practices for evaluation, management, and harm reduction. A socio-ecological public health framing will be used to examine interrelated and dynamic systems that impact the health of children and adults with ADHD including examining the scope of practice and education of NPs in mental health care competencies as well as access to quality care, bias, and stigma (see Fig. 2.1).

## 2.2 Scope of Practice and Education of NPs in Mental Health Care Competencies

Although mental health is appreciated as essential to health, the inclusion of sufficient mental health care education in nonpsychiatric NP programs varies greatly. Psychiatric Mental Health Nurse Practitioner (PMHNP) students are required to take advanced physical assessment, yet a comparable course on psychiatric assessment is not required for nonpsychiatric NP students. For unclear reasons, the psychiatric evaluation and mental status exam are not specifically included in the AANP standards for health assessment/evaluation (AANP 2022), and up until recently, the National Organization of Nurse Practitioner Faculties (NONPF) competencies for NPs were developed and published separately for each population served. Collaboration between the American Association of Colleges of Nursing and the NONPF, followed by the roll-out of the NONPF (2024) competency implementation guide for NP faculty, shifted the emphasis in education to shared competencies. The expectation is that the depth of focus on specific populations will differ between NP programs, but not the competencies themselves. The change also signals the realization that separation of mind and body in health and healthcare is somewhat arbitrary, and health is influenced by multiple social and ecological systems. The ten domains of NP competency are described in Table 2.1, along with examples of content that are discussed in this paper, and that might be useful to faculty for curriculum planning.

**Table 2.1** Foundations of ADHD curriculum content and competencies for nurse practitioners

| NONPF NP core competency domains | NONPF NP core competency descriptions | Content related to ADHD for NP curricula |
|---|---|---|
| Domain 1<br>Knowledge of practice | The NP integrates, translates, and applies established and evolving scientific knowledge from diverse sources as the basis for ethical clinical judgment, innovation, and diagnostic reasoning | Genetic and environmental risk factors<br>Neurodevelopment; brain structure and function; neurotransmitter systems and targets for treatment<br>Microbiota-immune-nervous system interactions |
| Domain 2<br>Person-centered care | The nurse practitioner uses evidence-based and best practices to design, manage, and evaluate comprehensive person-centered care that is within the regulatory and educational scope of practice. Fundamental to person-centered care is respect for diversity, differences, preferences, values, needs, resources, and determinants of health unique to the individual | Therapeutic relationship; trust<br>Screening tools<br>Comprehensive psychiatric evaluation, ROS, MSE<br>Formulation based on synthesis of multiple sources of data across intraindividual, interindividual, organizational, and societal systems<br>Differential diagnosis in the context of multiple comorbidities<br>Treatment planning based on patient factors and acuity<br>Use of clinical practice guidelines to plan and implement pharmacological and nonpharmacological parent training and whole person support, as indicated<br>Consideration of cultural diversity, preferences, values, resources |
| Domain 3<br>Population health | The nurse practitioner partners, across the care continuum, with public health, healthcare systems, community, academic community, governmental, and other entities to integrate foundational NP knowledge into culturally competent practices to increase health promotion and disease prevention strategies in effect the care of populations | Epidemiology of ADHD across the lifespan<br>Epidemiology and needs assessment in communities<br>Community partnerships<br>Implicit bias, health disparities and inequities<br>Leadership to promote principles of social justice, health equity |

(continued)

**Table 2.1** (continued)

| NONPF NP core competency domains | NONPF NP core competency descriptions | Content related to ADHD for NP curricula |
|---|---|---|
| Domain 4<br>Practice scholarship and translational science | The nurse practitioner generates, appraises, synthesizes, translates, integrates, and disseminates knowledge to improve person-centered health and systems of care | Current ADHD practice guidelines for youth and adults<br>Critical evaluation of literature<br>Translate evidence and apply in practice<br>Integrate evidence into practice<br>Evaluation of outcomes |
| Domain 5 quality and safety | The nurse practitioner utilizes knowledge and principles of translational and improvement science methodologies to improve quality and safety for providers, patients, populations, and systems of care | Harm reduction: Preventing diversion and overdose<br>Guidelines for telehealth visits<br>Culture of safety through quality improvement |
| Domain 6<br>Interprofessional collaboration in practice | The nurse practitioner collaborates with the interprofessional team to provide care through meaningful communication and active participation in person-centered and population-centered care | Interprofessional collaboration with team; patient advocacy; sensitivity to diverse organizations, cultures, and populations<br>Team-based care<br>Collaboration to develop, implement, and evaluate healthcare strategies |
| Domain 7<br>Health systems | The nurse practitioner demonstrates organizational and systems leadership to improve healthcare outcomes | Systems knowledge for working effectively across symptoms and the continuum of care<br>Structural and social determinants of health<br>Importance of integrated and collaborative PMH and PC systems |
| Domain 8<br>Technology and information literacy | The nurse practitioner envisions, appraises, and utilizes informatics and healthcare technologies to deliver care | Technologies (apps) for stress reduction, exercise, wellness, health literacy<br>Patient and caregiver needs, values, learning styles<br>Address gaps in access, knowledge, and information literacy<br>Quality, accountability, ethics, and cost-effectiveness |

(continued)

**Table 2.1** (continued)

| NONPF NP core competency domains | NONPF NP core competency descriptions | Content related to ADHD for NP curricula |
|---|---|---|
| Domain 9 Professional acumen | The nurse practitioner demonstrates the attributes and perspectives of the nursing profession and adherence to ethical principles while functioning as a committed equal partner of the interprofessional healthcare team | Culturally and linguistically appropriate care<br>Ethical principles in complex healthcare situations<br>How past experiences guide present and future practice<br>Strategies to prevent one's own personal biases from interfering with delivery of quality care<br>Accountability for practice within the regulatory standard and scope of educational preparation |
| Domain 10 Personal and professional leadership | The nurse practitioner participates in professional and personal growth activities to develop a sustainable progression toward professional and interpersonal maturity, improved resilience, and robust leadership capacity | Create an environment that promotes self-care, health, Well-being. Support whole person health and holistic Well-being of self<br>Applies principles of appreciative inquiry<br>Acknowledge potential and actual biases<br>Develop the capacity for leadership; mentor peers; engage in advocacy efforts to address health disparities, social justice, and equity to improve healthcare outcomes |

## 2.3 Socio-ecological Framings: The Individual

The person is the core of the socio-ecological framework. By extension, care of the whole person requires an understanding of what disorders a person has in addition to how multiple lifestyle and system influences contribute to exacerbation or resilience. The NONPF (2024) competency domains for NP education are discussed within the socio-ecological framework. Table 2.1 includes sample components of each of the ten domains that are relevant to ADHD educational content.

### 2.3.1 Knowledge of Practice: Scientific Evidence

The neuroscience relevant to ADHD and other neurodevelopmental disorders is making rapid advances, and in order to translate evidence into practice for care of patients with ADHD, NPs require the knowledge and skills to keep up with the ever-changing literature in pathophysiology, brain neuroanatomy and neurophysiology, and the circuitry that underlies functions. Recent evidence highlights the complexity of systems involved in neurodevelopment and neurodevelopmental disorders (e.g., Al-Haddad et al. 2019; Jakobi et al. 2024; Mulder et al. 2023, 2024; Warren et al. 2024).

***Neurodevelopment*** Of the mental disorders, ADHD is one of the most heritable, estimated at 74%; yet specific predictive genetic patterns remain elusive (Faraone and Larsson 2019). Interactions between genetic vulnerability and environmental insults at specific times and with sufficient exposure during fetal brain development are associated with ADHD and other neurodevelopmental and neuropsychiatric disorders. Several interacting systems are implicated: gut microbiota, immune, and the peripheral and central nervous systems. Neuroinflammation during early life with changes in structures and pathways is a leading hypothesis in the etiology and chronicity of ADHD.

***Neuroinflammation*** The gut microbiota are important regulators of immune function and neurodevelopmental processes in health and disease. Initial colonization of the gut begins during gestation through exposure to maternal microorganisms in utero and during vaginal delivery. Maternal dysbiosis from infectious processes, stress, diet, or treatments (e.g., antibiotics) during pregnancy can lead to release of proinflammatory cytokines and an inflammatory response that trigger the immune system of the fetus. Brain microglial cells, in addition to their role in CNS immune response, are important in tagging and clearing synapses for pruning, promoting neuronal wiring, and guiding neuronal differentiation and myelination. One of the current etiological theories for neurodevelopmental disorders is related to the impact of neuroinflammatory immune processes in utero and in early life that delay or impair prefrontal executive pathways and functions.

***Structures and Functions*** Structural differences between the brains of children with ADHD and children without ADHD demonstrate inefficient information processing in prefrontal circuits regulated by norepinephrine and dopamine. The prefrontal cortex develops over the first two decades, first with rapid growth of neurons, then with pruning for efficient information processing. The prefrontal cortex, together with its connections to the limbic system and basal ganglia, is essential for executive functioning and regulation of emotions and behaviors. Persistent symptoms of irritability, inattention, and dysregulated behaviors are shared across at least 20 Diagnostic and Statistical Manual of Mental Disorders-Fifth Edition (DSM-5) disorders that manifest early in life, suggesting a potential link between early neuroinflammation and white matter (oligodendrocyte) injury (Festa et al. 2025; Leibenluft et al. 2024; McKay et al. 2024).

### 2.3.2 Person-Centered Care

The core advanced physical assessment required for NP education includes some psychiatric assessment and diagnosis but most likely does not teach the critical interviewing and formulation skills that are useful for differentiating ADHD from other mental disorders with overlapping symptomatology, including disruptive mood dysregulation disorder (DMDD), oppositional defiant disorder (ODD), and bipolar, depressive, trauma, and anxiety disorders. Screening tools may help with determining the presence or absence of symptoms/behaviors, frequency, and intensity, but are not to be used in place of an evaluation.

***Psychiatric Interview and Evaluation*** The psychiatric evaluation is distinct in having very few objective diagnostic measures (labs, physical findings). Clinical guidelines that reduce the risk for misdiagnosis of ADHD include either an in-person or hybrid model of care with in-person evaluation. Because symptoms of ADHD are shared with numerous mental and physical disorders, and because the diagnosis requires corroboration of childhood symptoms, two pre-treatment consultations are recommended for adults who present with symptom complaints congruent with ADHD: the first for an in-person comprehensive mental health assessment and the second to review the material collected and to interview for corroboration of childhood symptoms (Hong et al. 2024).

Unlike the physical examination, the mental status exam (MSE) is primarily observational and occurs throughout the psychiatric interview. Depending upon the responses during the interview, the provider may also need to ask about mood (at the time of the visit) and ask about or clarify thought content (e.g., worries, beliefs, phobias, delusions, suicidal, or homicidal) and perceptions (e.g., hearing things that others don't hear). Many new to the MSE forget that it is an examination of the present status within the timeframe of the encounter; it is not about what the patient thought or did yesterday, or plans to do next week. Depending upon the age of the patient, the provider may or may not observe hyperactive or impulsive behaviors, or irritability. Whatever behaviors are observed are those that are recorded—and the rest is history. Gathering a comprehensive history to make a differential diagnosis takes at least 1 h with additional review, synthesis, and collaborative discussion and treatment planning with patients and, if indicated, their families or caregivers.

The timeline for development of symptoms is essential to the differential diagnosis. If the patient reports symptoms prior to the age of 12, validated self-report and observer (parent and teacher) screening measures should be completed and interpreted. For adults, corroboration of early symptoms should be obtained from school reports or family members to determine whether inattention and hyperactivity symptoms were present before age 12. Multiple screening tools are used to identify symptoms, symptom severity, and duration and to monitor response to treatment. The Connors scales, among others, are available for patient self-report as well as corroboration from teachers and caregivers. The Adult ADHD

Self-Report Scale (ASRS-5) is a useful screening and monitoring tool, with good sensitivity and specificity that is aligned with the American Psychiatric Association's (2022) Diagnostic and Statistical Manual of Mental Disorders (DSM-5-TR). ADHD diagnostic criteria incorporate examples relevant to adult life, such as difficulties managing finances or maintaining focus at work (APA 2022). If the patient reports symptoms congruent with ADHD but onset after age 12, or as an adult, reports symptoms prior to age 12 that cannot be corroborated, then a diagnosis of ADHD is generally not supported, and nonpharmacological approaches may be more appropriate.

***ADHD Comorbidities*** A differential diagnosis for ADHD relies on careful assessment of all potential psychiatric and medical contributors or comorbidities. Whereas children with untreated ADHD are likely to display hyperactive or impulsive behaviors, adults are more likely to complain of inattention and to have a greater number of comorbidities. ADHD shares symptoms with, and is often comorbid with, multiple mental disorders characterized by emotional and behavioral dysregulation, including anxiety disorders, major depressive disorder, bipolar disorder, and substance use disorders, so a comprehensive history, physical and psychiatric evaluation, and review of systems and symptoms are required for a differential diagnosis (Choi et al. 2022; Substance Abuse and Mental Health Services Administration 2021). Medical conditions that may mimic ADHD include thyroid dysfunction, vitamin deficiencies, sleep disorders, and neurological issues. Screen for other substance use such as alcohol, caffeine, or recreational drugs, as well as medications or chronic health conditions, on ADHD-like symptoms. To aid in the differential diagnosis, many validated mood, anxiety, and substance use scales can be helpful for data gathering. Assessment of adult patients with suspected drug misuse behaviors may include a drug urine screen.

***Diagnostic Formulation and Differential Diagnosis*** Formulation models are useful for synthesizing data gathered during the psychiatric evaluation in a way that guides treatment recommendations. A diagnosis of ADHD is made when the patient meets all criteria required by the American Psychiatric Association's *Diagnostic and Statistical Manual for Mental Disorders* (DSM-5-TR, 2022), but it does not inform providers about other contributing factors that may be important foci to target in the treatment plan. Two formulation models that are often used in psychiatry are the four *Perspectives of Psychiatry* (McHugh and Slavney 1998) and the biopsychosocial 4Ps (Warren et al. 2024). The four perspectives are what the patient **H**as, **I**s, **D**oes, and **E**xperiences or has experienced (**HIDE**), and the four Ps are the **p**resenting, **p**redisposing, **p**recipitating, and **p**erpetuating factors. The socio-ecological approach is also a model used to identify interventions at the levels of the individual, relationship, community, and society levels. The socio-ecological Whole School, Whole Community, Whole Child (WSCC, Centers for Disease Control and Prevention 2024) framework is useful for formulation and treatment planning for children with ADHD, because the school is often the place where functional attentional and behavioral problems are first noticed, and school teachers and mental health care providers are collaborative treatment partners.

***Pharmacological Treatment*** Effective treatment of ADHD depends upon an accurate diagnosis with symptom onset by age 12 and clinically significant interference with functioning, assessment of current and previous treatments and their indications and efficacy, and the patients/families preferences. Best practices guidelines and consensus statements on treatment of ADHD are readily available and updated periodically. They are generally considered current for 5 years. National organizations that regularly evaluate and publish guidelines for diagnosis and treatment of ADHD include the American Academy of Child and Adolescent Psychiatry, American Academy of Pediatrics, American Psychiatric Association, and the Veterans Administration. Consensus statements that reflect larger groups include the Association of European Psychiatrists and the World Federation of ADHD, among others.

Pharmacological approaches are available to children through adulthood, starting at age 3. Stimulant formulations, consisting of amphetamine and methylphenidate preparations are generally the first line of treatment. Stimulants, at the appropriate dose, are generally well tolerated, improve outcomes, and reduce all-cause mortality (Li et al. 2024; US Department of Veterans Affairs 2022). Immediate release stimulants have a rapid onset but may also lose effectiveness within 4–6 h. Extended-release formulations last through most of the school or work day but may interfere with sleep. Youth 6–17 years of age with ADHD that are not getting sufficient restorative sleep may benefit from the one of the approved alpha2-adrenergic agents as a monotherapy for ADHD or for augmentation with a morning stimulant.

If a diagnosis of ADHD is supported and comorbid disorders are also present, the World Health Federation (Faraone et al. 2021) recommends that treatment proceed in order of priority based upon safety and severity, even if that means holding off on treatment for ADHD. Persons with comorbid substance use disorders may be at risk for misuse, overdose, or diversion of stimulants, and stimulants may initially worsen symptoms of comorbid anxiety disorders, so should be stabilized prior to initiating stimulants.

***Harm Reduction*** The FDA (2022) makes the following recommendations as part of the boxed warning for stimulants: (1) Before prescribing, assess each patient's risk for abuse, misuse, and addiction. (2) Educate patients and their families about the risks, proper storage of the drug, and proper disposal of any unused drug. (3) Throughout treatment, reassess each patient's risk, and frequently monitor for signs and symptoms of abuse, misuse, and addiction. Patients should be educated about the risks and how to store and dispose of unused drugs. Throughout treatment, patients should be reassessed and monitored for signs and symptoms of drug misuse or use disorder. Reporting by the National Institute of Drug Abuse (NIDA 2024) indicates that overdose deaths from stimulants more than quadrupled between 2015 and 2022, with 70% of deaths in 2022 involving fentanyl.

Harm reduction strategies for at-risk populations (e.g., those with substance use histories or disorders; young adults that might divert or share stimulants) include use of stimulant alternatives, extended-release stimulants, and prescription and education around using naloxone should overdose be suspected. In general, extended-release stimulants are safer than short-acting stimulants. Education about the risks, regular monitoring, and support are essential. Equipping patients with an opioid reversal drug, such as naloxone nasal spray or auto-injection, and verifying understanding of rationale reduce risks for harm.

***Nonpharmacological Treatment*** To provide person centered care for ADHD consistent with current best practice guidelines, psychotherapeutic competencies could be strengthened in NP education. Several evidence-based psychotherapeutic strategies may be implemented by NPs with additional training, certification, and expertise to deliver them. Multiple organizations offer training and certification in parent training to nonpsychiatric clinicians. Parent training is an 8- to 10-week psychotherapeutic intervention that is aimed at helping parents provide structure and positive support for behavioral regulation in the home (see CDC 2024, May). Older children and adolescents may benefit from digital tools that support emotional regulation and well-being (see CDC n.d., Digital Tools Fact Sheet). Older youth and adults with persistent irritability and low frustration tolerance may benefit from psychoeducation and guidance in understanding triggers to irritability and using coping skills. The Veterans Administration Circle of (whole) Health is an interactive tool with resources for adults to identify and address other lifestyle variables that may be contributing to emotional dysregulation (e.g., sleep, diet) and inattention.

Development of collaborative relationships with psychotherapists improves organizational whole person care capacity. Evidence-based psychotherapies such as cognitive behavioral therapy are provided by psychotherapists who have been professionally trained and licensed to treat mental, emotional, and behavioral disorders by psychological means. Psychotherapy is a core competency of PMHNPs, social workers, professional counselors, psychologists, and psychiatrists. ADHD is a chronic disorder, and psychotherapy is an important option for patients during challenging periods in life.

## 2.4 Socio-ecological Framings: Interdependent Strategies to Advance Health Equity

### 2.4.1 Access to Quality Care

Mental health services cannot be separated from underlying sociocultural and economic structures that inform assurance of health equity—the absence of unfair, avoidable, or remediable differences in health and well-being among groups of people (Lazo-Porras and Penniecook 2023). Understanding how these unique contextual experiences are intertwined with the daily lived experiences of individuals diagnosed with ADHD is key to how they are potentially linked to access quality mental health services, health insurance, and ensuring the provision of culturally relevant care. Recognizing key areas shaping access to quality care for ADHD including the cost of care, untrustworthiness of healthcare providers and healthcare institutions, and reliable and accessible transportation including the location of services will be discussed (Lazo-Porras and Penniecook 2023). This is critical since timely and quality healthcare can be considered as a basic human right, and good access to these services for ADHD can contribute to improvements in a person's health and well-being (Ma et al. 2017). This is particularly poignant for marginalized individuals, racialized Black and brown populations, and stigmatized groups facing difficulties with gaining access to quality care (Alves-Bradford et al. 2020). Knowing that access to quality care is generally outside an individual's control, socio-ecological framing is used to provide a holistic lens to assess a person's access to quality care for ADHD considering multilevel factors—intrapersonal, interpersonal, institutional, community, and policy levels. Recognizing the influence of various contexts, systems, and environments on individuals and families is essential for developing a comprehensive understanding of a problem and subsequently providing effective support for healthy development and functioning.

### 2.4.2 Intrapersonal and Interpersonal Levels: Equity

A person's attitudes, beliefs, knowledge, and behaviors about access to quality care for ADHD is reliant on the clarity of understanding about the disorder and knowledge (ADHD literacy) including sufficient information about the benefits of treatment to pursue care. Moreover, it is important to understand the person's finances which can delay access to quality care for ADHD along with perception of the severity of ADHD symptoms. To access care, the price of services must be considered in relation to the income of those seeking care. In addition, knowledge about ADHD among family and the social network is significant since their perceptions about it inform the advice they will share about accessing quality care. Social support is an important determinant of pursuing care. Their close circle of family and peers can also serve as a source of emotional and practical support that can influence the person's access to quality care for ADHD. On account of cultural norms, religiousness or spirituality as well as stigma and shame must also be considered (Breland-Noble et al. 2015; DuPont-Reyes et al. 2020; Rice and Harris 2021). These factors are all shaped by the person's developmental level, positionality, and life experience which interacts with institutional, community, and policy-level factors (e.g., policies and legislation related to race, ethnicity, immigration status, and language) (Alves-Bradford et al. 2020). These are not neutral, and today they drive many of the health inequities regarding mental health outcomes therefore they serve as a backdrop against which individuals navigate their daily lives (Lazo-Porras and Penniecook 2023).

### 2.4.3 Community/Organizational Level: Equity

Social values and perception of ADHD among the community influence a person's motivation to access quality care; however, this too is dependent upon individual characteristics including education, income, and household circumstances. The location of the community and the availability of services affect accessibility to quality care. This includes considering the volume (coverage) and type of existing services to meet the needs of individuals diagnosed with ADHD (Woodward et al. 2024). Often, access to quality care is designated by geographical service location; however, this is too basic and problematic. Accessibility to care is shaped by the relationship between the location of the services and the location of those diagnosed with ADHD (e.g., transportation, travel time, distance, and cost) (Woodward et al. 2024). Services available on a county or state level can be sporadically dispersed, concentrated in some areas compared to others (urban versus suburban versus rural) with differential resources. Thus, mental health provider density, including provider competence and skills in ADHD, culturally relevant practices to effectively care for populations from diverse backgrounds, and acceptance of insurance to initiate care, can produce unfavorable odds for those of lower economic status, requirements of language-matched care services, and racialized Black and brown populations (Alves-Bradford et al. 2020). Luelmo et al. (2020) state that access to diagnostic and intervention services is not equal, with the most vulnerable families and children having limited access. Chung et al. (2024) assert that disparities in access to care for Black populations is posited to persist throughout the multi-stage process of ADHD service use (e.g., recognition of ADHD, accessing and engaging with a mental health provider, cultural and relational values, and other structural barriers). Accessibility is also driven by the relationship of attitudes of those diagnosed with ADHD about personal and practice qualities of services to the real qualities of the existing services, including providers' attitudes about acceptable personal attributes of those seeking care for ADHD (Woodward et al. 2024).

Institutions also play a critical role in a person's access to quality care for ADHD. Moreover, relationships and communication between institutions such as schools, work settings, and health institutions can influence accessibility to care (Alves-Bradford et al. 2020). Available and affordable culturally relevant care informs access to quality ADHD services, and these interact with both intra- and interpersonal factors. This is reflected by how approachable providers and institutions are regarding their knowledge about ADHD and transparent information that is readily available on services offered (e.g., culturally specific care) as well as the acceptability of care which is driven by trustworthiness felt by the populations being cared for. Negative experiences of people diagnosed with ADHD using professional mental health services within institutions can stem from provider bias and discrimination and this reinforces the need for culturally responsive training in order to provide care that is culturally relevant (Alves-Bradford et al. 2020). The relationship between the organization of resources including appointment systems used, hours of operation, walk-in facilities, as well as the capability of individuals diagnosed with ADHD to accommodate these factors must be considered (Woodward et al. 2024). Additionally, when institutions want to demonstrate access to quality care for ADHD, they should establish practices such as outreach about services offered including their affordability, waiting time for initiation, and ensure appropriate coordination between services (Coleman et al. 2022). These intentional steps limit social marginalization and stigmatization and promote an environment of being welcomed and respected (Yansen et al. 2025) (Table 2.2).

### 2.4.4 Societal and Policy Level: Equity

Policy-level factors inform regulations that affect all other levels, especially access to quality care at the institutional level. Addressing community-level factors necessitates policy changes, advocacy efforts, and the provision of culturally responsive and affordable ADHD care. Naturally, changes in individual and community attitudes cannot be viewed in isolation from policy and funding. It is crucial to enhance mental literacy related to ADHD at school, primary care offices, and other contexts. It is also important to recognize the intersectionality of race, age, socio-economic status, and gender in addition to the intersectionality of inequitable care systems and work toward equitable and accessible quality ADHD care for all in need (Lazo-Porras and Penniecook 2023). Policies often overlook socioeconomic inequalities where out-of-pocket payment is required for accessing care which can lead to inequitable mental health outcomes exacerbating ADHD symptoms and functional impairment. Thus, accessibility directly influences early diagnosis and culturally relevant treatment of ADHD. Financial support for groups that are made to be vulnerable through policy decisions must be attended to for assurance of equity in mental health care to prevent critical delays. For example, the integration and expansion of telehealth and other digital care solutions is critical especially in rural areas with limited access to ADHD care. Policies must also prioritize initiatives that promote diversity within the mental health workforce including the recruitment and retention of those with specialized ADHD skills. Given the impact of policy-level actions on all other levels, coordinated actions between policy especially funding

**Table 2.2** Access to care factors

| |
|---|
| Awareness that services exist |
| Acceptability of services |
| Availability and accommodation of services |
| Affordability of services |
| Appropriateness of services |

mechanisms for insurance, awareness initiatives about ADHD, anti-stigma campaigns, and mitigation of other barriers are advised to ensure quality care for impacted individuals.

## 2.5 Socio-ecological Framings: Approaches to Building Awareness and Reducing Implicit Bias and Stigma

In striving to advance equity, utilizing a socio-ecological framing allows for exploration of major determinants that affect mental health across the lifespan. For under-resourced populations diagnosed with ADHD, these factors can lead to mental health inequalities—differences in the health of individuals or groups. Moreover, under-resourced groups tend to have several intersecting social adversities, and having ADHD as an additive concern, inequity becomes a leading contributor to poor mental health (Ani et al. 2024). What's more, institutional power structures that marginalize populations through practices that are not inclusive must also be cognizant of implicit bias demonstrated by providers which can lead to clinical inertia—providers fail to treat ADHD among some groups. Both implicit bias and stigma are commonly associated with mental health disorders including ADHD.

Implicit bias stems from attitudes or stereotypes that may be unintentional but influences a person's understanding, actions, and decisions impacting receipt of equitable care for ADHD (McCleary-Gaddy and Scales 2019). This can have enduring implications within our society since every person has implicit bias through how they were raised regarding perceptions of mental health concerns such as ADHD as well as characteristics of populations impacted by the disorder. Likewise, stigmas are negative beliefs held in society which encompass elements of stereotyping, prejudice, and discrimination that unfairly target certain groups. Stigma also connotes a mark of shame and humiliation through which the uniqueness of the shamed individual becomes lost (Nguyen and Hinshaw 2020). ADHD stigma therefore relates to the devalued social identity a person has due to the negative attribute of this mental disorder. Implicit bias is commonplace; however, it is critical to recognize and manage effectively to not perpetuate harm onto others via implicit bias and stigma. In the realm of mental healthcare, implicit bias and stigma impact the quality of care a person receives as well as their overall well-being (Ani et al. 2024). Further examination of implicit bias and stigma concerning ADHD for under-resourced populations will occur in the subsequent section applying a socio-ecological framing.

### 2.5.1 Intraindividual and Interpersonal Levels: Stigma

Interpersonal or social stigma can be layered affecting a person's intrapersonal perspective because of differences labeled as undesirable that violate the preferred norms of dominance. Values of a culture can also adversely impact a person because of discernable and indiscernible characteristics, including skin color, language, gender, as well as ADHD (Goffman 1963). These labeled differences tend to be used to discriminate against and exclude people, reducing their status, opportunities, and equity in society. What's more, the USA's culture and conventional media promotes messaging that suggests that people with mental health concerns including ADHD are incompetent, helpless, disinterested, rude, unreliable, immature, and responsible or at-fault for their illness. Conceivably, by repetitively combining ADHD with undesirable attributes, stigmatized attitudes surface against people with ADHD perceiving them as blameworthy and lacking intellect.

Intrapersonal or self-stigma is the internalization of toxic messages and beliefs about oneself (Nguyen and Hinshaw 2020). This becomes harmful when the person diagnosed with ADHD becomes aware of social stigma about ADHD, agrees with social stigma, and applies it to their own experience (Woodgate et al. 2020). Taken together, the action-oriented application process is when harm can inflict deleterious consequences for the person diagnosed with ADHD including low self-esteem, low self-efficacy, low self-confidence, as well as self-discrimination which can lead to isolation and contribute to later adverse social outcomes (Nguyen and Hinshaw 2020; Woodgate et al. 2020). For marginalized and historically disadvantaged groups, this social stigma can be layered, and one can view this through Dr. Patricia Hill Collin's work titled the matrix of domination (Hill-Collins 2009).

Interpersonal stigma can also flow from parent/guardian to child. These adults who are critical treatment agents in the process may have stigma about ADHD that instigates unwillingness to advocate for or pursue ADHD evaluation and treatment (Alleyne et al. 2025). Perceived public stigma about ADHD may produce fear that a diagnostic label will alter how their child is treated and perceived. As such, this child would have fewer opportunities and lower expectations toward success with personal endeavors. Guardians/parents were also noted to embody courtesy stigma—worry about being judged negatively for having a child diagnosed with ADHD (Alleyne et al. 2025). In addition, worry extended to medication stigma with concern about addiction and personality change. It is critical that recognition of compounding stigmas intersect with other identities, intensifying concerns about marginalization (Alleyne et al. 2025).

Countering stigmatization related to ADHD is a cognitive skill that should improve with development. One's reaction and response to these elements can be countered by stigma opposition and self-protection through a myriad of practices. These practices can include ignoring or challenging the stereotypes/stigma, combating internalized stigma via mental health (ADHD) literacy and education, as well as increasing contact with individuals affected by the disorder (Mannarini et al. 2023). Kooij et al. (2019) also described how young people were poised to combat and challenge ADHD stigma by self-disclosure and openness about their condition. In addition, stigma was lowered in teachers toward individuals impacted by ADHD when they had greater knowledge about the condition (Kooij et al. 2019).

### 2.5.2 Intraindividual and Interpersonal Levels: Implicit Bias

When considering implicit bias toward self, individuals diagnosed with ADHD are required to exercise deep thinking of one's internalized bias and an acknowledgement of the impact of ADHD on their personhood. Engaging in consciousness-raising practices through daily critical reflection, sharpened awareness about ADHD including culturally, and developmentally appropriate knowledge can guide a change in perception and perspective about ADHD. Contextualizing an internalized negative way of thinking about ADHD is important considering the larger social dynamics at play, especially for marginalized groups interfacing with oppressive practices and ideologies which helps to reframe narratives about self and build personal resilience. Intrapersonal bias has also surfaced with the model minority myth. Internalized beliefs that Asian Americans are independently high achieving without psychological distress can perpetuate a false narrative (Park and Foley-Nicpon 2022). Underlying this notion is the belief that low academic performance and distress in Asian Americans stem from inferior traits or behaviors of the child. This bias is associated with the under-identification of ADHD in academic venues (Chung et al. 2024).

Implicit bias from an interpersonal stance is revealed in a myriad of contexts from school and teacher interaction to the diagnostic process with clinicians. Lack of cultural awareness of symptom expression and cultural norms could lead to bias when clinicians make cross-cultural diagnostic judgments using the DSM-5-TR. It is essential for clinicians to attend to both their own implicit biases and blind spots during the treatment process (Cervantes et al. 2024). How a clinician engages patients and understands the dynamic forces at work along with experiences of people from diverse cultural groups influences approaches to defying the judgment and bias produced by their own personal norms, values, and beliefs (Cervantes et al. 2024).

Implicit bias also surfaces along the lines of essentialism—social categories viewed as reflecting a person's intrinsic essences or biological natures (He et al. 2024). The DSM-5-TR unequivocally indicates a connection between innate characteristics and the manifestation of alleged symptoms according to sex. This designation is reinforced by creating a distinction between sex and gender yet exclaiming nothing about gender related factors. Moreover, there is no explicit discussion about gender in the section titled Culture Related Diagnostic Issues (Koutsoklenis and Honkasilta 2023). As an alternative, diagnostic matters related to sex and gender are diminished to biopsychological qualities and assumptions related to female/male binary (or sex) (Koutsoklenis and Honkasilta 2023). This illustrates essentialism since ADHD is described as having a fixed essence, for example, genetic, neurodevelopmental dysfunction, because of contrasts in binary sexes. This totally negates the socially constructed roles, behaviors, expressions, and identities associated with gender pluralism in addition to disregarding how sociocultural aspects such as gender role and cis normativity are entwined with psychosocial characteristics that may be revealed as behaviors considered symptoms (Koutsoklenis and Honkasilta 2023).

In addition, when using an inclusive lens, bias can lead to underdetection of ADHD via mislabeling symptoms as oppositional or disruptive in marginalized socially oppressed ethnic and racialized populations. This can be produced because of explicit or implicit clinician bias (Koutsoklenis and Honkasilta 2023). Considering this, it is critical for clinicians to employ culturally competent practices in diagnosing members of socially oppressed ethnic or racialized groups to ensure correct diagnosis (Koutsoklenis and Honkasilta 2023). Similarly, Alleyne et al. (2025) found that minoritized children and families in the USA who are fluent in languages other than English are less likely to be diagnosed with ADHD and begin or continue care. While barriers are numerous including provider implicit bias (e.g., racial, ethnic, language), historical medical discrimination, lack of culturally congruent care, and stigma, it is requisite that inequities in care are reduced and intentional proactive efforts are made to build rapport with all individuals and families.

### 2.5.3 Community/Organizational and Policy Level: Stigma

Organizational stigma related to ADHD surfaces when people affected by the disorder experience stigma within institutions or organizations during daily actions in the community including public and private establishments ranging from school to work to healthcare. Organizational stigma can have a great impact on help-seeking for a person diagnosed with ADHD (Zhang et al. 2021). Moreover, it is generally sensed that this stigma is controllable because it characteristically follows organizational players who actively elect to be involved in stigmatizing those with ADHD. Therefore, change is possible when those in power within these organizations act as supporters and advocates for equity-oriented measures for those diagnosed with ADHD. This moves beyond individual efforts to manage stigma to foster more systemic change (Zhang et al. 2021). Similarly, community stigma is when prejudices or negative attitudes associated with ADHD occur, and this can be manifested through judgment and gossip in the local community. Fear of community stigma can spur distress and if assistance is sought the individual diagnosed with ADHD may seek care outside of their immediate location.

Stigma on the broader level is rooted in societal policies, laws, and barriers that serve to further marginalize people from stigmatized groups (e.g., ADHD labeling, racial and ethnic identity, non-English speaking). Power differentials strongly inform the experience of a person diagnosed with ADHD considering the social, physical, and economic resources aligned with a person's status and influence to enforce societal norms to favor themselves and subjugate those who deviate from these norms (McCleary-Gaddy and Scales 2019).

Policies that are not intentional with reducing ADHD stigma can be challenged by groups with strong, persistent opposition that provides a clear counter narrative that reveals the full humanity of the persona. They can also challenge the way the stigma is represented (McCleary-Gaddy and Scales 2019). National organizations and federal agencies such as the Mental Health Association and the Substance Abuse and Mental Health Services Administration have both constructively crafted anti-stigma campaigns to transform the culture and perception of how individuals with mental health concerns such as ADHD, including substance use disorders, are perceived by society. Many other organizations can join in to lead the way such as the American Psychiatric Nurses Association, the International Society for Psychiatric Nurses, the Joint Commission on the Accreditation of Healthcare Organizations, and the American Psychological Association.

### 2.5.4 Community/Organizational and Policy Level: Implicit Bias

Community-level factors associated with bias are more locally concentrated and relate to environments that are proximal to the person affected by ADHD. For example, Alleyne et al. (2025) argue that responses parents note within their communities about their child's ADHD symptoms may reinforce concerns related to bias. A double standard exists with behaviors demonstrated by racialized Black and brown children where there are more negative racial attitudes with disruptive behavior (Kang and Harvey 2020). Parents may receive criticisms about their child and worry about biased perceptions and inequity. As we live and work in spaces that sustain biased views and perpetuate racism, people in society must continually examine and challenge their own biases including clinicians before, in, and after the clinical encounters (Alleyne et al. 2025).

Organizational bias functions within institutions or systems of power and creates inequitable policies or practices that result in certain social groups being favored and others being disadvantaged. Psychology is an institution of knowledge which directs practice and guidelines for mental health disorders which inform policy. This institution of knowledge is influenced by the medical model of disability which has traditionally centered a deficit framework of clinical diagnosis. Moreover, it conceptualizes the person deemed to have a disability as negatively different from the standard, normal, or average person (Darazsdi 2022). Consequently, the medical model establishment pathologized divergence from a supposed normal individual and posits relative significance on conformity as much as possible to a socially constructed norm (Darazsdi 2022). When we reflect on bias, the alleged normal is a nebulous and ever-changing concept that is created from a set of values that serve the agenda of the dominant group, and the institution of mental health care has sustained these values by limiting its concepts of illness to these societal standards. In the medical space, in the field of psychology and in the USA, non-disabled white men set these norms, and the standards are productivity, capitalism, and Eurocentrism (Darazsdi 2022). Accordingly, society creates subjective and limiting standards for human behavior, rooted in the beliefs of one privileged portion of the population (Darazsdi 2022). While ADHD is a serious and chronic disorder, establishing a more inclusive lens on what normal represents will require deep reflection on philosophical views and value judgments since there is no biological test for ADHD.

From a policy stance, zero tolerance-policies contribute to bias toward Black children. They receive harsher punishment compared to white children even when the same behaviors are demonstrated. Black students lose twice as many school days to suspensions or expulsions by comparison, even when students were chastised for comparable offenses. Moreover, racialized Black students were disciplined most often for subjective offenses including non-compliance or for showing disrespect at three times the rates for white students (Behnken 2014). Similarly, Haley (2021) found a link between the application of policy, race, children's behaviors, and school discipline. When schools perceive children as troublemakers and those children interfere with the learning opportunities for all students, then disciplinary measures are justified. However, the actual differences noted in behavioral problems between Black or Latin X and white students were found to be an effect of the racial disparities in school discipline—behavior of Black and Latino children with identical misbehavior as their white counterparts received harsher consequences—this is racial discrimination at play. Moreover, when a Black or LatinX student is deemed disobedient or bad, there's limited perception of this behavior stemming from an underlying medical condition but instead is deemed to be linked to a fundamental deficiency. By attributing the source of behavior problems contrarily, schools carry out a bifurcated practice of responding to students racialized as Black and brown compared to white students. Policies may seem neutral on the surface but when activated in practice bias occurs because people make judgments based on their values, socialized norms, and inequitable practices.

## 2.6 Conclusions

Access to quality care for ADHD can be significantly influenced by multiple systems (Fig. 2.1). Sociopolitical constructs such as race and ethnicity are also drivers of access to quality care for ADHD through socio-structural/policy, institutional, interpersonal, and internal mechanisms. Differential access to quality care must be mitigated to eliminate inequality and promote health equity for those impacted by ADHD. Greater depth of psychiatric and mental health care education for NPs also has the potential to strengthen NP knowledge, skills, and attitudes for expanded holistic care of persons with ADHD across NP practice settings. The NONPF (2024) competency Domains 9 and 10: Professional Acumen and Leadership emphasize reflective practice to advocate for community, organizational, and societal changes that reduce health inequities and support the well-being of patients and families.

## References

Al-Haddad BJS, Oler E, Armistead B, Elsayed NA, Weinberger DR, Bernier R, Burd I, Kapur R, Jacobsson B, Wang C, Mysorekar I, Rajagopal L, Adams Waldorf KM. The fetal origins of mental illness. Am J Obstet Gynecol. 2019;221(6):549–62. https://doi.org/10.1016/j.ajog.2019.06.013.

Alleyne K, Fogler J, Simon KM, Spinks-Franklin A. Promoting equitable engagement in attention-deficit/hyperactivity disorder care. Pediatr Ann. 2025;54(1):e12–7. https://doi.org/10.3928/19382359-20241007-01.

Alves-Bradford JM, Trinh NH, Bath E, Coombs A, Mangurian C. Mental health equity in the twenty-first century: setting the stage. Psychiatr Clin North Am. 2020;43(3):415–28. https://doi.org/10.1016/j.psc.2020.05.001.

American Academy of Child and Adolescent Psychiatry (AACAP) Committee on Community-Based Systems of Care and AACAP Committee on Quality Issues. Clinical update: child and adolescent behavioral health care in community systems of care. J Am Acad Child Adolesc Psychiatry. 2024;62(4):367–84.

American Association of Nurse Practitioners. Standards of practice for nurse practitioners: discussion paper; 2022. https://www.aanp.org/advocacy/advocacy-resource/position-statements/standards-of-practice-for-nurse-practitioners.

American Psychiatric Association. Diagnostic and statistical manual of mental disorders. 5th ed; 2022. https://doi.org/10.1176/appi.books.9780890425787.

Ani C, Ola B, Hodes M, Eapen V. Editorial: equity, diversity and inclusion in child and adolescent mental health—equality of opportunities should be every child's right and every society's obligation. Child Adolesc Ment Health. 2024;29:123–5. https://doi.org/10.1111/camh.12698.

Behnken M. Linking ADHD to incarceration among African Americans. ADHD Rep. 2014;22(7):9–16.

Breland-Noble AM, Wong MJ, Childers T, Hankerson S, Sotomayor J. Spirituality and religious coping in African-American youth with depressive illness. Ment Health Relig Cult. 2015;18(5):330–41.

Centers for Disease Control and Prevention. Parent training in behavior management for ADHD; 2024. https://www.cdc.gov/adhd/treatment/behavior-therapy.html

Centers for Disease Control and Prevention. Digital tools for supporting emotional well-being in children and teens; n.d. https://nap.nationalacademies.org/resource/other/dbasse/wellbeing-tools/interactive/pdf

Cervantes J, Wallace J, Helms AK, Diekroger EA, Fogler J. Complex ADHD challenging case: when simple becomes complex: managing clinician bias and navigating challenging family dynamics in a 6-year-old girl with ADHD and developmental delays. J Dev Behav Pediatr. 2024;45(2):e181–4. https://doi.org/10.1097/DBP.0000000000001265.

Chung S, Williams A, Owens E, McBurnett K, Hinshaw SP, Pfiffner LJ. Parental cognitions, treatment engagement, and child outcomes of ADHD behavioral treatment among Asian American families. Res Child Adolesc Psychopath. 2024;52(3):325–37. https://doi.org/10.1007/s10802-023-01139-9.

Choi W-S, Woo Y-S, Wang S-M, Lim H-K, Bahk W-M. The prevalence of psychiatric comorbidities in adult ADHD compared with non-ADHD populations: A systematic literature review. PLoS ONE. 2022;17(11):e0277175. (PLOS)

Coleman M, Cuesta-Briand B, Collins N. Rethinking accessibility in light of the Orange Declaration: applying a socio-ecological lens to rural mental health commissioning. Front Psych. 2022;13:930188. https://doi.org/10.3389/fpsyt.2022.930188.

Danielson ML, Claussen AH, Bitsko RH, Katz SM, Newsome K, Blumberg SJ, Kogan MD, Ghandour R. ADHD prevalence among U.S. children and adolescents in 2022: diagnosis, severity, co-occurring disorders, and treatment. J Clin Child Adolesc Psychol. 2024;53(3):343–60. https://doi.org/10.1080/15374416.2024.2335625.

Darazsdi ZC. Examining the relationship between autistic people's self-esteem and perceived bias in their alliance with their mental health practitioner; 2022. Retrieved from https://gt-law.idm.oclc.org/login?url=https://www.proquest.com/dissertations-theses/examining-relationship-between-autistic-peoples/docview/2679792008/se-2

DuPont-Reyes MJ, Villatoro AP, Phelan JC, Painter K, Link BG. Adolescent views of mental illness stigma: an intersectional lens. Am J Orthopsychiatry. 2020;90:201–11. https://doi.org/10.1037/ort0000425.

Faraone SV, Larsson H. Genetics of attention deficit hyperactivity disorder. Molecular Psychiatry, 2019;24(4):562–575. (Nature)

Faraone SV, Banaschewski T, Coghill D, Zheng Y, Biederman J, Bellgrove MA, Newcorn JH, Gignac M, Al Saud NM, Manor I, Rohde LA, Yang L, Cortese S, Almagor D, Stein MA,

Albatti TH, Aljoudi HF, Alqahtani MMJ, Asherson P, Atwoli L, et al. The world federation of ADHD international consensus statement: 208 evidence-based conclusions about the disorder. Neurosci Biobehav Rev. 2021;128:789–818. https://doi.org/10.1016/j.neubiorev.2021.01.022.

Festa LK, Jordan-Sciutto KL, Grinspan JB. Neuroinflammation: an oligodendrocentric view. Glia. 2025; https://doi.org/10.1002/glia.70007.

Goffman E. Stigma: Notes on the management of spoiled identity. Prentice-Hall; 1963.

Haley J. Disparities in discipline: An interview with Jayanti Owens. Brown Political Review; 2021 https://brownpoliticalreview.org/disparities-in-discipline-bpr-interview-jayanti-owens/.

He J, Zhao Y, Zhang C, Wang W, Huang J. Relationship between ethnic essentialism and psychological compatibility: Roles of ethnic identity and self-construal. Behavioral Sciences. 2024;14(12):1207. https://doi.org/10.3390/bs14121207.

Hill-Collins P. The new politics of community. Am Sociol Rev. 2009;75(1):7–30. https://doi.org/10.1177/0003122410363293.

Hong J, Mattingly G W, Carbray J A, Cooper T V, Findling R L, Gignac M, Glaser P E, Lopez F.A, Maletic V, McIntyre R S, Robb A S, Singh M K, Stein M A, & Stahl S M. (2024). Expert consensus statement for telepsychiatry and attention-defi cit hyperactivity disorder. CNS Spectrums, 29(5), 1–12. https://doi.org/10.1017/S1092852924000208.

Jakobi B, Cimetti C, Mulder D, Vlaming P, Franke B, Hoogman M, Arias-Vasquez A. The role of diet and the gut microbiota in reactive aggression and adult ADHD-an exploratory analysis. Nutrients. 2024;16(14):2174. https://doi.org/10.3390/nu16142174.

Kang S, Harvey EA. Racial differences between black parents' and white teachers' perceptions of attention-deficit/hyperactivity disorder behavior. J Abnorm Child Psychol. 2020;48:661–72. https://doi.org/10.1007/s10802-019-00600-y.

Kooij JJS, Bijlenga D, Salerno L, Jaeschke R, Bitter I, Balázs J, Thome J, Dom G, Kasper S, Nunes Filipe C, Stes S, Mohr P, Leppämäki S, Casas M, Bobes J, Mccarthy JM, Richarte V, Kjems Philipsen A, Pehlivanidis A, Niemela A, et al. Updated european consensus statement on diagnosis and treatment of adult ADHD. Eur Psychiatry. 2019;56:14–34. https://doi.org/10.1016/j.eurpsy.2018.11.001.

Koutsoklenis A, Honkasilta J. ADHD in the DSM-5-TR: what has changed and what has not. Front Psych. 2023;13:1064141. https://doi.org/10.3389/fpsyt.2022.1064141.

Lazo-Porras M, Penniecook T. Health equity: access to quality services and caring for underserved populations. Health Policy Plan. 2023;38(Supplement_2):ii1–2. https://doi.org/10.1093/heapol/czad073.

Leibenluft E, Allen LE, Althoff RR, Brotman MA, Burke JD, Carlson GA, Dickstein DP, Dougherty LR, Evans SC, Kircanski K, Klein DN, Malone EP, Mazefsky CA, Nigg J, Perlman SB, Pine DS, Roy AK, Salum GA, Shakeshaft A, Silver J, et al. Irritability in youths: a critical integrative review. Am J Psychiatry. 2024;181(4):275–90. https://doi.org/10.1176/appi.ajp.20230256.

Li L, Zhu N, Zhang L, et al. ADHD pharmacotherapy and mortality in individuals with ADHD. JAMA. 2024;331(10):850–60. https://doi.org/10.1001/jama.2024.0851.

Luelmo P, Sandoval Y, Kasari C. Undocumented Mexican mothers of children with autism: navigating the health care and educational service systems. Int J Dev Disabil. 2020;68(4):567–77. https://doi.org/10.1080/20473869.2020.1850159.

Ma PHX, Chan ZCY, Loke AY. The socio-ecological model approach to understanding barriers and facilitators to the accessing of health services by sex workers: a systematic review. AIDS Behav. 2017;21(8):2412–38. https://doi.org/10.1007/s10461-017-1818-2.

Mannarini S, Taccini F, Rossi AA. Stigma toward internalizing and externalizing disorders: how do adolescents perceive their peers? A network analysis approach. J Res Adolesc. 2023;33(3):803–15. https://doi.org/10.1111/jora.12839.

McCleary-Gaddy AT, Scales R. Addressing mental illness stigma, implicit bias, and stereotypes in medical school. Acad Psychiatry. 2019;43:512–5. https://doi.org/10.1007/s40596-019-01081-3.

McHugh PR, Slavney PR. The perspectives of psychiatry. 2nd ed. Baltimore: Johns Hopkins University Press; 1998.

McKay CC, Scheinberg B, Xu EP, Kircanski K, Pine DS, Brotman MA, Leibenluft E, Linke JO. Modeling shared and specific variances of irritability, inattention, and hyperactivity yields novel insights into white matter perturbations. J Am Acad Child Adolesc Psychiatry. 2024;63(12):1239–50. https://doi.org/10.1016/j.jaac.2024.02.010.

Mulder D, Aarts E, Arias Vasquez A, Bloemendaal M. A systematic review exploring the association between the human gut microbiota and brain connectivity in health and disease. Mol Psychiatry. 2023;28(12):5037–61. https://doi.org/10.1038/s41380-023-02146-4.

Mulder D, Jakobi B, Shi Y, Mulders P, Kist JD, Collard RM, Vrijsen JN, van Eijndhoven P, Tendolkar I, Bloemendaal M, Arias Vasquez A. Gut microbiota composition links to variation in functional domains across psychiatric disorders. Brain Behav Immun. 2024;120:275–87. https://doi.org/10.1016/j.bbi.2024.05.037.

National Institute of Drug Abuse (NIDA). US overdose deaths: facts and figures; 2024.. https://nida.nih.gov/research-topics/trends-statistics/overdose-death-rates#Fig 6

National Organization of Nurse Practitioner Faculties. Nurse practitioner role core competencies; 2024.

Nguyen P, Hinshaw S. Understanding the stigma associated with ADHD: Hope for the future? ADHD Rep. 2020;28(5):1–22.

Park S, Foley-Nicpon M. Excellence expected, needs overlooked: implications for working with Asian American twice-exceptional students. Teach Except Child. 2022. https://doi.org/10.1177/00400599221097020.

Rice AN, Harris SC. Issues of cultural competence in mental health care. J Am Pharm Assoc. 2021;61(1):e65–8. https://doi.org/10.1016/j.japh.2020.10.01.

Staley BS, Robinson LR, Claussen AH, Katz SM, Danielson ML, Summers AD, Farr SL, Blumberg SJ, Tinker SC. Attention-deficit/hyperactivity disorder: diagnosis, treatment, and telehealth use in adults. National Center for Health Statistics Rapid Surveys System, United States. 2023. MMWR Morb Mortal Wkly Rep. 2024;73(40):890–5. https://doi.org/10.15585/mmwr.mm7340a1.

Substance Abuse and Mental Health Services Administration. Addressing stigma associated with mental illness and substance use disorders; 2021.

U.S. Department of Veterans Affairs. Identification and management of attention-deficit/hyperactivity disorder (ADHD) in adults: academic detailing clinician guide; 2022. https://www.pbm.va.gov/PBM/Academic-DetailingService/Documents/508/10-1520_ADHD_ClinicianGuide_P97040.pdf

U.S. Food and Drug Administration. FDA requires updates to improve safe use of prescription stimulants used to treat ADHD and other conditions; 2022.

Warren M, O'Connor C, Lee JE, Burton J, Walton D, Keathley J, Wammes M, Osuch E. Predispose, precipitate, perpetuate, and protect: how diet and the gut influence mental health in emerging adulthood. Front Nutr. 2024;11:1339269. https://doi.org/10.3389/fnut.2024.1339269.

Woodgate RL, Comaskey B, Tennent P, Wener P, Altman G. The wicked problem of stigma for youth living with anxiety. Qual Health Res. 2020;30(10):1491–502. https://doi.org/10.1177/1049732320916460.

Woodward A, Fuhr DC, Barry AS, Balabanova D, Sondorp E, Dieleman MA, Pratley P, Schoenberger SF, McKee M, Ilkkursun Z, Acarturk C, Burchert S, Knaevelsrud C, Brown FL, Steen F, Spaaij J, Morina N, de Graaff AM, Sijbrandij M, Cuijpers P, et al. Health system responsiveness to the mental health needs of Syrian refugees: mixed-methods rapid appraisals in eight host countries in Europe and the Middle East. Open Res Eur. 2024;3:14. https://doi.org/10.12688/openreseurope.15293.

Yansen AR, Muttaqin MZ, Prasetya YY. Towards democratised psychiatry: building metacommunities for inclusive and equitable global mental health. BJPsych Int. 2025;22(1):8–10. https://doi.org/10.1192/bji.2024.30.

Zhang R, Wang MS, Toubiana M, Greenwood R. Stigma beyond levels: advancing research on stigmatization. Acad Manag Ann. 2021;15(1):88–222. https://doi.org/10.5465/annals.2019.0031.

# Comprehensive Assessment of ADHD Across the Lifespan

# 3

Katiah Llerena and Angela Melland

## 3.1 Guidelines and Expert Consensus for Evaluation of ADHD: Strengths and Limitations

The evaluation and diagnosis of attention-deficit/hyperactivity disorder (ADHD) in the United States primarily follow guidelines established by the Wolraich et al. (2019) and the Society for Developmental and Behavioral Pediatrics (2020). It is imperative to understand the strengths and limitations of the existing US guidelines, highlighting concerns such as their reliance on outdated diagnostic criteria, their limited scope in addressing adult ADHD, and their narrow inclusion of age-related data.

The 2019 AAP guidelines serve as a widely accepted framework for healthcare providers, offering a standardized approach to ADHD diagnosis and treatment in children and adolescents. Their strengths lie in their emphasis on multi-informant assessments, incorporating information from parents, teachers, and standardized rating scales to enhance diagnostic accuracy (AAP 2019). Additionally, the AAP guidelines advocate for evidence-based behavioral treatments as a first-line approach, particularly for younger children, before considering pharmacological options (Wolraich et al. 2019). Their accessibility ensures widespread utilization in pediatric settings across the country, making ADHD evaluations relatively standardized. Despite these strengths, the guidelines rely heavily on outdated DSM-IV criteria as the foundation rather than the DSM-5-TR (American Psychiatric Association

K. Llerena (✉)
Department of Psychiatry & Behavioral Sciences–TrPR and THRIVE,
University of California, San Francisco, CA, USA
e-mail: katiah.llerena@ucsf.edu

A. Melland
Psychiatric Mental Health Nurse Practitioner-Board Certified (PMHNP–BC),
University of Minnesota, Minneapolis, MN, USA

© The Author(s), under exclusive license to Springer Nature Switzerland AG 2026
R. J. Kennedy (ed.), *Assessment & Treatment of ADHD*,
https://doi.org/10.1007/978-3-032-24171-9_3

2022) updates, which better account for ADHD in adults. They also focus solely on individuals aged 4–18, leaving a significant gap in guidance for adults. ADHD frequently co-occurs with other mental health conditions. Although a seventh Key Action Statement was added to address co-occurring conditions, the recommendations lack the depth offered by other guidelines, which provide a more comprehensive and multidisciplinary framework for evaluating and treating complex comorbid presentations (Eom and Kim 2023). Additionally, the AAP framework mostly relies on pediatric primary care providers but does not sufficiently address the need for multidisciplinary collaboration involving psychologists, psychiatrists, and other specialists.

Similarly, the SDBP (2020) guidelines provide valuable insight into ADHD, particularly in children and adolescents with complex presentations involving psychiatric, developmental, and medical comorbidities (Barbaresi et al. 2020). These guidelines acknowledge the variability in ADHD presentation and emphasize a multidisciplinary approach to assessment and treatment, promoting collaboration among specialties (Barbaresi et al. 2020). However, they do not extend their recommendations to adults. Additionally, while the SDBP guidelines offer a more nuanced approach to ADHD assessment, their influence on broader healthcare policies remains relatively limited.

In stark contrast to guidelines from other countries, the USA remains the only developed nation without formalized national guidelines for adult ADHD. For a summary of international guidelines, see Table 3.1. To address the longstanding absence of national guidelines for adult ADHD, the American Professional Society for ADHD and Related Disorders (APSARD), in partnership with Children and Adolescents with ADHD (CHADD), announced in January 2025 that it is developing the first US guidelines dedicated to adult ADHD, with an anticipated release in late 2025. This long-overdue initiative aims to establish standardized diagnostic and treatment protocols for adults with ADHD. The forthcoming guidelines are expected to provide a national standard for adult ADHD care, incorporate a lifespan perspective, offer guidance for individuals receiving a late diagnosis, and present a comprehensive approach to managing comorbidities.

**Table 3.1** Summary of international guidelines

| Country/ organization | Guideline status | Key features |
|---|---|---|
| World Federation of ADHD | Global consensus statement on adult ADHD (2021) | 208 evidence-based conclusions<br>International scope<br>Lifespan perspective on ADHD |
| Australia (AADPA) | AADPA Guidelines (2022) | 113 clinical recommendations across lifespan<br>Multimodal treatment approach<br>Focus on everyday functioning and quality of life<br>Addresses cultural and geographic considerations |

(continued)

**Table 3.1** (continued)

| Country/ organization | Guideline status | Key features |
|---|---|---|
| United Kingdom (NICE) | NICE Guidelines (2018, updated 2019) | Covers children, young people, and adults<br>Detailed recommendations for pharmacological and psychosocial interventions<br>Emphasis on recognition, diagnosis, and management<br>Includes monitoring and review of medication |
| Germany | Integrated adult ADHD into national guidelines (2018) | Continuity of care focus<br>Adult ADHD inclusion in national framework |
| Spain | Integrated adult ADHD into national guidelines (2017) | Emphasizes adult ADHD diagnosis and management in national standards |
| Canada (CADDRA) | CADDRA Guidelines (2018) | Covers ages 6–18 years<br>Diagnostic criteria and treatment recommendations<br>Provider guidance<br>Recommends structured clinical interviews |
| United States (AAP) | Clinical Practice Guideline (2019) | Covers ages 4–18 years<br>Seven Key Action Statements (KAS)<br>Chronic care model approach<br>Limited guidance on comorbidities |

## 3.2 Components of a Comprehensive Diagnostic Evaluation for ADHD

A comprehensive diagnostic evaluation for ADHD involves a detailed process that aims to understand the individual's symptomatology, functional impairments, developmental history, and relevant contextual factors. This evaluation goes beyond merely identifying symptoms that fulfill ADHD criteria; it encompasses an assessment of the broader psychosocial, medical, and cultural contexts from which these symptoms emerge (Barkley 2023; Wolraich et al. 2019). The goals of a comprehensive evaluation include accurate diagnosis, identification of comorbid conditions, and formulation of individualized treatment plans. A robust evaluation is essential to avoid misdiagnosis and inappropriate treatment, which can exacerbate impairments and perpetuate stigma (Willcutt 2012). See Table 3.2 for an overview of the key components of a comprehensive ADHD diagnostic evaluation.

Nurse practitioners (NPs)—especially those with psychiatric mental health training—are well-equipped to lead comprehensive ADHD evaluations. Their holistic training enables them to assess not only psychiatric and behavioral symptoms but also physical health, lifestyle factors, and social determinants that impact symptom presentation and management. As frontline providers, NPs often play a central role in identifying concerns, initiating assessments, and coordinating care across multidisciplinary teams.

**Table 3.2** Key components of a comprehensive ADHD diagnostic evaluation

| Assessment component | Purpose | Populations/settings |
|---|---|---|
| Initial screening tools | Identify individuals with and at risk and guide further evaluation | All ages, primary care, behavioral health, school settings |
| Clinical interview | Gather developmental, psychosocial, academic, and symptom history | All ages, primary care, behavioral health |
| Parent/caregiver/partner input | Contextualize behavior across settings and history | All ages |
| Rating scales | Quantify symptom severity and frequency | Primary care, behavioral health, school, specialty |
| Structured interviews | Increase diagnostic reliability using DSM-5 criteria | Child/adult variants in specialty settings |
| Multi-informant reports | Cross-validate symptoms in different environments | School, home, occupational contexts |
| Functional impairment evaluation | Understand day-to-day impact across domains | School, work, home contexts |
| Medical evaluation | Rule out physical or systemic contributors | Pediatric/adult primary care, psychiatric care |
| Comorbidity screening | Identify mood, anxiety, substance use, and learning disorders | Behavioral health and integrated care |
| Neuropsychological testing and objective measures | Assess cognitive, executive, and attentional functioning | Complex or ambiguous presentations, behavioral health |
| Strength-based/neurodiverse tools | Highlight strengths, sensory profiles, and adaptive strategies | Neurodivergent-friendly clinical settings |

It is important to note that the ADHD assessment process conducted by NPs is rarely performed in isolation. NPs often collaborate with other medical providers, psychologists, psychiatrists, educators, specialists, and other mental health professionals to ensure a comprehensive evaluation. This multidisciplinary approach may involve consulting about medical history, reviewing educational or neuropsychological assessments, conferring with treating therapists, and coordinating with schools to support appropriate accommodations. By integrating insights from various professionals, NPs contribute their unique perspective while helping to build a thorough and informed understanding of the patient's diagnoses and needs.

The initial screening and referral process is a critical first step in the comprehensive evaluation of ADHD. Screening often begins in primary care or educational settings, with primary care providers often being at the frontline for ADHD screening and initial evaluation. For children, early developmental or educational concerns—such as language delays or learning challenges—often indicate further assessment (Wilens et al. 2003). Adults may seek diagnosis from a mental health or primary care provider due to functional impairments in academics, work, home, and/or interpersonal relationships. Aside from general ADHD symptoms, referral indicators include persistent symptoms that interfere with daily functioning, poor response to behavioral or academic supports, presence of comorbid psychiatric or medical conditions, developmental concerns, diagnostic ambiguity, or inadequate response to medication. Providers may consider using brief, validated screening tools designed to identify individuals at risk for ADHD (Wolraich et al. 2019). These referral indicators should be thoroughly documented to ensure clarity regarding the rationale for referral and to facilitate the coordination of care.

### 3.2.1 Comprehensive Clinical Interview

The comprehensive clinical diagnostic interview is a cornerstone of ADHD assessment and remains one of the most widely used techniques in evaluating psychological conditions (Herbert et al. 2010; Mueller and Segal 2014). This assessment begins with setting clear expectations, including a discussion of confidentiality, the purpose of the visit, and the possibility that a diagnosis—or medication prescription—may not occur immediately (Adamou et al. 2024). After obtaining consent, the provider explores the presenting concern, often involving difficulties at school, work, or home.

The primary goals are to identify symptom patterns that align with ADHD and obtain a comprehensive patient history, while building rapport. NPs should adapt their interviewing style to support patients who struggle with attention, including using a chronological structure or redirecting when necessary. Throughout the clinical interview, the provider gathers detailed developmental, social, educational, occupational, medical, and psychiatric histories while simultaneously observing for hallmark ADHD features in line with the DSM-5-TR criteria. For children, this includes evaluating whether at least six symptoms of inattention and/or hyperactivity-impulsivity have been present for a minimum of 6 months; for adults, five symptoms are required. The DSM-5-TR identifies three ADHD presentations—predominantly inattentive, predominantly hyperactive-impulsive, and combined—each with distinct symptom profiles. NPs carefully assess the presence of symptoms as well as their severity, frequency, and impact on functioning across settings. While unstructured clinical interviews are commonly used to gather the necessary diagnostic information, they allow considerable variability depending on the provider's theoretical orientation, training, and personal biases, which may limit reliability (Mueller and Segal 2014).

To improve diagnostic consistency, particularly in complex cases, NPs may opt to use or refer to structured or semi-structured diagnostic interviews. These tools provide an evidence-based framework to systematically assess symptom onset, persistence, and cross-situational impairment, enhancing diagnostic reliability (Silverstein et al. 2020). Structured interviews, such as the Structured Clinical Interview for DSM-5 (SCID-5; (First et al. 2015)) for adults or the Kiddie Schedule for Affective Disorders and Schizophrenia (K-SADS; (Kaufman et al. 1997) for children, are particularly valuable when differentiating ADHD from overlapping conditions such as anxiety, depression, trauma, autism spectrum disorders, sleep disorders, or learning disorders.

Regardless of the method, a thorough ADHD assessment should adopt a biopsychosocial framework that considers social determinants of health, cultural background, family dynamics, trauma exposure, and resilience factors to support accurate diagnosis and culturally responsive care (Trivedi and Stiebel 2020). A thorough psychiatric review, along with input from multiple informants, structured rating scales, and sometimes objective performance measures, supports accurate diagnostic formulation. Ultimately, the NP must work within their scope of practice. If diagnostic uncertainty remains—or if complex comorbidities are present—referral to an ADHD specialist, psychiatrist, or neuropsychologist for further evaluation is recommended to ensure diagnostic accuracy and optimal patient outcomes.

### 3.2.2 Screening and Rating Scales

Rating scales are essential tools in the assessment of ADHD. These measures offer objective, norm-referenced data and are designed to evaluate the frequency and severity of ADHD symptoms and, in some cases, associated functional impairments across various domains, such as attention, behavior regulation, and executive functioning. While these measures do not replace the need for a clinical interview to determine diagnosis, they are invaluable in screening for ADHD, quantifying symptom presence, supporting diagnostic formulation, and monitoring symptom severity and treatment progress. For example, the ASRS v1.1 is a psychometrically robust self-report scale that can be used to screen for ADHD and track symptom severity to indicate treatment progress. Other common screening and rating scales are presented in Table 3.3. NPs should select tools that are developmentally appropriate, validated for the target population, and sensitive to cultural and linguistic differences. The most common screening tools and rating scales used by NPs are summarized in Table 3.3. All listed measures demonstrate strong psychometric properties (Caroline S. et al. 2024; Mulraney et al. n.d.).

### 3.2.3 Multi-informant Assessment Approach

A multi-informant approach is essential to ADHD assessment, as diagnosis requires evidence of symptoms and impairment across multiple settings (American Psychiatric Association 2022). Input from parents, teachers, caregivers, and—in adult cases—partners or workplace supervisors provides a well-rounded picture of functioning. These perspectives can be gathered through interviews or standardized rating scales (Table 3.3) to help contextualize behaviors across environments. For instance, teachers may observe inattention or classroom disruption, while parents might note emotional reactivity or struggles with routines at home. In adults, workplace informants may comment on time management, productivity, attention to detail, or interpersonal dynamics.

**Table 3.3** Screening and rating scales

| Measure | Target population | Format | Administration time | Key features | Clinical utility for NPs |
|---|---|---|---|---|---|
| Adult ADHD Self-Report Scale (ASRS v1.1) | Adults (18+) | Self-report questionnaire | 5 min | Based on WHO criteria<br>Part A items most predictive of ADHD<br>Consistent with DSM-5-TR criteria<br>Specifically worded for adult symptom presentation<br>Symptoms over the last 6 months | Brief screening tool ideal for time-limited appointmentsTracks symptom severity over time |
| Vanderbilt ADHD Diagnostic Rating Scale (VADRS) | Children (6–12) | Parent and teacher rating forms | 10–15 min | Includes all 18 DSM criteria for ADHD<br>Screens for common comorbidities<br>Assesses academic performance | Comprehensive assessment tool<br>Multi-informant input enhances diagnostic accuracy<br>Shorter follow-up versions available for monitoring treatment |
| Conners Rating Scale | Children, adolescents, and adults | Self-report and observer forms | 5–20 min | Widely used in clinical and research settings<br>Assesses multiple domains (hyperactivity, impulsivity, emotional problems) | Can be used to monitor treatment effectiveness<br>Indicates whether symptoms affect daily functioning |
| Conners Adult ADHD Rating Scales 2nd Ed. (CAARS-2 | Adults (18+) | Self-report and observer forms | Full: 10–20 min<br>Short: 5–10 min<br>Index: 1–3 min | Extended age norms (up to 70+)<br>Comprehensive ADHD assessment<br>Multiple versions for different clinical needs | Multiple formats offer flexibility based on clinical setting and time constraints<br>Useful for diagnosis, treatment planning, and progress monitoring |

(continued)

**Table 3.3** (continued)

| Measure | Target population | Format | Administration time | Key features | Clinical utility for NPs |
|---|---|---|---|---|---|
| Wender Utah Rating Scale (WURS-25) | Adults (18+) | Self-report retrospective assessment | 10–15 min | Retrospectively assesses childhood ADHD symptoms<br>Selected for discriminative ability in distinguishing adults with/without ADHD | Addresses DSM requirement for childhood onset<br>Particularly useful when childhood records unavailable<br>Complementary to current symptom measures |
| Swanson, Nolan and Pelham Teacher and Parent Rating Scale (SNAP-IV) | Children and adolescents | Parent/teacher rating scale | 5–10 min | Based on DSM criteria<br>Measures both inattention and hyperactivity-impulsivity domains | Brief and easy to administer<br>Validated across diverse populations<br>Useful for monitoring symptom changes |

It is important to note that informant discrepancies are common and diagnostically valuable, often reflecting environmental differences, cultural expectations, or unrecognized scaffolding that masks impairment in some contexts (Silverstein et al. 2020). Rather than defaulting to the most severe or most frequently reported symptoms, clinicians should thoughtfully synthesize all perspectives (Martel et al. 2021). Additionally, developmental stage is key when selecting informants, as adolescents with late-onset ADHD may exhibit unique features—such as metacognitive deficits, trauma exposure, or high parental pressure—that standard informants may overlook (Sibley et al. 2022). Therefore, it is important to integrate informant input from multiple sources to enhance diagnostic accuracy and support individualized care.

### 3.2.4 Functional Impairment Scales

Assessing functional impairment within the developmental context is a critical component of a comprehensive ADHD evaluation. While the presence of symptoms is necessary for diagnosis, it is not sufficient without evidence that these symptoms interfere with daily functioning. Importantly, recent literature suggests that true remission in ADHD should encompass not just symptom resolution but also functional recovery (Weiss 2022). While functional impairment can be gathered during the clinical interview, there is a growing emphasis on using validated scales (Sasser et al. 2016). Table 3.4 lists psychometrically sound ADHD-focused functional impairment tools. For children, the WFIRS-P is a leading choice for detailed, domain-specific assessment, and the BFIS (with parent input) offers a norm-referenced overview of impairment. For adolescents, NPs should utilize both adolescent self-report and parent-report functional scales, such as the WFIRS-S and WFIRS-P combination (Weiss 2022). Additionally, cross-cultural validation studies have demonstrated that the WFIRS-S performs consistently across different countries and languages (Weiss 2022). The BFIS also offers a norm-based gauge of chronic impairment severity in teens (Weiss et al. 2018). The IRS remains an option for a quick global rating.

Current practices in assessing functional impairment in adults with ADHD are quite variable. Practitioners use a "diverse range" of tools, sometimes repurposing instruments not ideally suited for ADHD (Fuermaier et al. 2024). NPs are encouraged to use structured impairment scales—particularly WFIRS-S for its ADHD focus and sensitivity and the BFIS for its normative benchmarking—alongside clinical diagnostic interviews and symptom checklists. Beyond facilitating a comprehensive evaluation of adult ADHD and accurate diagnosis, these tools also provide valuable metrics for tracking treatment effectiveness beyond symptom reduction. Of note, functional data should be interpreted through a developmental lens, considering age-appropriate expectations, environmental demands, and the buffering effects of individual strengths and supports.

**Table 3.4** Functional impairment scales

| Tool name | Target age group(s) | Domains assessed | Timeframe for rating | Best use case |
|---|---|---|---|---|
| Weiss functional impairment rating scale–parent report (WFIRS-P) | Children (parent-report) | Family, school, life skills, self-concept, social activities, risky behaviors | Past month | Detailed assessment of ADHD-specific impairment in children |
| Weiss Functional Impairment Rating Scale-Self Report (WFIRS-S) | Adolescents, adults (self-report) | Work, family, social life, self-concept, risk behaviors, daily activities | Past month | Monitoring ADHD-related impairment in teens and adults |
| Barkley Functional Impairment Scale (BFIS) | Children, adolescents, adults (norm-referenced) | 15 domains: academic, occupational, social, financial, legal, etc. | Past 6 months | Normative comparison of impairment severity across lifespan |
| Brown Executive Function/ Attention Scales (Brown EF/A Scales) | Children (3–7 years), School Age (8–12 years), Adolescent (13–18 years), and Adult (19+ years). Self, Parent & Teacher Report | Executive functioning and ADHD related impairments including: activation/ task initiation, attention and focus, effort/sustained attention, emotional regulation, working memory, and self-monitoring | 6 months | Functional impairments in academic, occupational, and everyday life settings. International Translation of Language. Persons with High IQ, Comorbidities. |
| ADHD-FX Scale | Children | Home, school, peers (mixed symptoms and function) | Not clearly defined | Preliminary research or screening; not ideal for clinical tracking |
| Sheehan Disability Scale (SDS) | Adults | Work/school, social, family (broad disability) | Current functioning | Very brief screen in adult psychiatry |
| WHO Disability Assessment Schedule (WHODAS 2.0) | Adults, older adults | Cognition, mobility, self-care, getting along, life activities, participation | 30 days | Generic disability measure, especially in older adults |

### 3.2.5 Neuropsychological and Objective Measures

Neuropsychological and objective performance-based measures can be valuable tools in the assessment of ADHD, especially in complex or ambiguous cases. NPs may initiate referrals for such testing or interpret results from evaluations previously conducted. These assessments are designed to evaluate cognitive domains frequently impacted by ADHD, such as attentional control, working memory, processing speed, and executive functioning (Barkley 2011). One important class of objective measures includes computerized tasks such as the Sustained Attention to Response Task (SART; (Robertson et al. 1997) and Continuous Performance Tests

(CPTs), which offer quantitative, standardized metrics of attention, impulsivity, and cognitive control. These tools provide a valuable complement to subjective rating scales and clinical interviews, adding an empirical layer of data that may reduce bias and increase diagnostic confidence—especially when self-report and informant reports are inconsistent or incomplete.

NPs should consider referring a patient for formal neuropsychological testing when the case is complex, the diagnosis is unclear, cognitive deficits are suspected, there is a poor response to treatment, or when objective documentation of cognitive and functional status is required for broader management or accommodations. Testing can also help detect symptom exaggeration. Neuropsychological tests can identify areas of cognitive strength—such as strong verbal reasoning, visual-spatial skills, or pattern recognition—that may be leveraged to support individuals in academic, occupational, or social contexts (Antshel et al. 2013). In addition, targeted assessment of working memory through short-story recall measures from the Children's Memory Scale (CMS) or the Wechsler Memory Scale (WMS), as well as Digit Span subtests from the Wechsler Intelligence Scale for Children (WISC) or Wechsler Adult Intelligence Scale (WAIS), can detect auditory short-term and working memory deficits that are common among adolescents and adults with ADHD, including those with average, above-average, and very high intellectual ability (Kennedy et al. 2019). Identifying these specific cognitive weaknesses alongside areas of strength provides valuable information for diagnosis, treatment planning, academic accommodations, and the development of individualized compensatory strategies. Recognizing these strengths enables providers to tailor recommendations, identify effective compensatory strategies, and guide accommodations that can meaningfully enhance educational and occupational outcomes.

NPs should be cautious when interpreting these tools in isolation. Performance can be influenced by situational variables such as fatigue, test anxiety, motivation, or a quiet testing environment that does not reflect real-world distractions (Toplak et al. 2013). Additionally, some individuals with ADHD may perform well in structured tasks under optimal conditions, even while struggling with similar demands in daily life. For this reason, neuropsychological and computerized tests are not diagnostic, and they are best used as part of a multimodal assessment strategy.

### 3.2.6 Associated Features Beyond the DSM-5

Although not part of the core DSM-5-TR diagnostic criteria for ADHD, a range of associated symptoms and features—such as emotional dysregulation, rejection sensitivity, and sensory over-responsivity—are commonly reported by individuals with ADHD and can significantly impact daily functioning. NPs should be aware of these non-core symptoms, as they can provide valuable insights into the individual's lived experience and functional impairments. Recognizing these features can inform more personalized treatment planning and may help explain variability in treatment response and long-term outcomes.

Emotional dysregulation is one of the most prominent associated symptoms, characterized by low frustration tolerance, irritability, and rapid mood shifts. These experiences are linked to difficulties across academic, occupational, and interpersonal settings (Shaw et al. 2014). The Difficulties in Emotion Regulation Scale (DERS) has been shown to correlate with ADHD symptoms and predicts functional impairment in adults, offering a valuable assessment tool in clinical settings (Gratz and Roemer 2004). Another commonly reported feature is rejection sensitive dysphoria (RSD), characterized by overwhelming emotional pain in response to perceived criticism or rejection, often resulting in either social withdrawal or disproportionate expressions of anger (Surman et al. 2013). Related to this, self-concept issues, including negative self-esteem and poor self-perception, are frequently self-reported by adults with ADHD and have been linked to higher levels of functional impairment (Chamberlain et al. 2016; Faraone and Kuntsi 2008; Szuromi et al. 2013).

Cognitive impulsivity, which includes racing thoughts, disorganized thinking, and hasty decision-making, can impair executive functioning and make it harder for individuals to engage effectively in treatment (Nigg et al. 2005). Additionally, sensory over-responsivity—manifesting as hypersensitivity to auditory, tactile, or visual input—is frequently observed and may overlap with conditions such as anxiety and ASD (Panagiotidi et al. 2018; Pfeiffer et al. 2015). Sluggish cognitive tempo (SCT)—characterized by daydreaming, mental fogginess, and slow processing speed—often co-occurs with ADHD, especially the inattentive presentation (Burns and Becker 2021).

Because standard ADHD measures often fail to capture these associated symptoms, NPs should consider routinely inquiring about these associated features in more depth. Longitudinal research and case studies continue to support the notion that associated features emotional and sensory symptoms are associated with long-term impairment and reduced treatment responsiveness in ADHD (DeSerisy et al. 2019; Grossman and Avital 2023; Shaw et al. 2014; Soler-Gutiérrez et al. n.d.). Comorbidity screening tools, including DSM-5 Cross-Cutting Symptom Measures, are particularly useful for identifying co-occurring conditions that often accompany emotional and sensory symptoms (Panagiotidi et al. 2018; Shaw et al. 2014). Some patient-reported measures, like the Barkley Quick-Check, are designed to capture non-core symptoms such as affective impulsivity and frustration intolerance and have demonstrated validity in adult populations (Barkley and Fischer 2010; Nigg et al. 2005).

### 3.2.7 A Neurodiverse-Affirming ADHD Assessment Approach

A neurodiverse-affirming approach to ADHD assessment challenges the traditional, deficit-based medical model by recognizing ADHD not as a disorder but as a difference in cognitive style, sensory processing, and executive functioning. This strengths-oriented framework acknowledges that individuals with ADHD experience and engage with the world in diverse ways that often diverge from neurotypical norms. Rather than pathologizing these differences, it emphasizes adaptive functioning, resilience, and support needs (Kapp et al. 2013). Providers are encouraged to adopt this perspective in qualitative interviews to explore patient narratives that can elucidate areas of strength such as hyperfocus, creativity, divergent thinking, or sensory attunement—qualities often overlooked in standardized diagnostic protocols. Positively reframing ADHD traits and inquiring about ADHD-related personal strengths (e.g., "ADHD Superpowers") can buffer against RSD and improve self-esteem (Müller et al. 2024). An affirming approach helps NPs build rapport, validate neurodivergent identities, and guide effective interventions, thus advancing equitable and compassionate care for individuals with ADHD.

### 3.2.8 Developmental, Gender, and Cultural Considerations in ADHD Assessment

It is important to tailor ADHD assessment to the individual's developmental stage because symptom presentation and functional impact differ markedly across the lifespan. In childhood, hyperactivity and impulsivity are typically more prominent, while in adolescence and adulthood, symptoms often shift toward inattention, executive dysfunction, and emotional dysregulation (Barkley 2011). Young children often lack the capacity to accurately self-report, making caregiver and teacher input essential at this point. As children grow, they may resist parental involvement, and internalized symptoms such as disorganization or academic underperformance may emerge. In adulthood, challenges like time management, sustaining focus in the workplace, and maintaining interpersonal relationships become more salient (Adler and Shaw 2011; Kooij et al. 2019). These adult presentations are frequently overlooked, especially when the history of childhood symptoms is incomplete or difficult to verify.

In addition to developmental stages, cultural and gender-related factors significantly influence the expression and recognition of ADHD symptoms. Cultural norms around behavior, communication styles, and academic expectations shape how symptoms are perceived and reported by families, educators, and clinicians. For example, Chan et al. (2022) found that parents in Hong Kong, due to high expectations for conformity and achievement, report ADHD symptoms at lower thresholds than UK parents (Chan et al. 2022). Although Hong Kong children are objectively less active, they are rated as more hyperactive and inattentive—likely reflecting higher parental stress and stricter norms. In contrast, Western parents may view higher activity levels as acceptable before considering them problematic. These patterns highlight the importance of culturally sensitive assessment practices.

Gender also plays a key role in ADHD diagnosis and presentation. Males are more frequently diagnosed, historically due to more overt hyperactive and disruptive behaviors that prompt referral (Stibbe et al. 2020). Girls, by contrast, more often present with inattentive symptoms—such as daydreaming, disorganization, and difficulty sustaining attention—which are less disruptive and more likely to go unnoticed. Implicit biases among educators and clinicians may further contribute to lower referral rates for girls, even when symptom severity is comparable.

In practice, nurse practitioners can apply gender and culturally informed strategies to improve diagnostic accuracy and build stronger therapeutic alliances. When working with females, it is important to probe for internalizing symptoms—such as perfectionism, rumination, and rejection sensitivity—and to explore compensatory behaviors like excessive list-making or mimicking others to mask difficulties. Questions such as, "Do you often feel mentally exhausted after trying to stay focused or organized all day?" can help uncover hidden impairments. Similarly, culturally responsive assessment begins by asking how a client's family or community views attention and behavior concerns. NPs should avoid pathologizing behaviors that may be adaptive in a specific context and normalize the assessment process to reduce stigma. When language barriers exist, using trained interpreters and culturally adapted assessment tools can further support equitable, accurate evaluation.

## 3.3 Functional Impairments Across Developmental Stages

Functional impairments in ADHD extend beyond core symptoms of inattention, hyperactivity, and impulsivity, affecting major life domains across developmental stages. These impairments persist from childhood into adulthood and vary depending on age, comorbidities, and environmental demands. Although formal measures for assessing functional impairment are described in Sect. 3.2.4, this section focuses on key behavioral indicators that NPs should assess during clinical evaluations.

### 3.3.1 Academic and Occupational Functioning

Academic challenges are among the earliest functional impairments observed. In childhood, excessive activity, impulsivity, cognitive inflexibility, and difficulties with inhibitory control contribute to problems completing tasks, sustaining attention, and adapting to academic demands (Harpin 2005; Sadozai et al. 2024). Research consistently shows that students with ADHD are at higher risk of underachievement, disciplinary actions, absenteeism, and school dropout compared to peers (French et al. 2024; Jangmo et al. 2019). Despite occasional areas of academic or occupational overachievement, these successes do not preclude a diagnosis of ADHD, as individuals may simultaneously underperform in other domains (Adamou et al. 2024).

As adolescents transition to adulthood, the nature of academic and occupational impairments evolves. Adolescents with ADHD often struggle more than their peers due to increased autonomy, reduced structure, and heightened demands (Solomon and Sparrow 2024). Difficulties launching into adulthood are common, with many experiencing lower rates of college attendance, increased dropout rates, higher unemployment, and lower-income employment compared to peers without ADHD (Howard et al. 2016; Katzman et al. 2017). Environmental transitions, such as moving away from family or starting employment, can exacerbate impairments due to executive function deficits.

### 3.3.2 Family Dynamics and Household Functioning

Understanding the influence of family dynamics and household functioning is critical to capturing the full context in which ADHD symptoms emerge and are managed. NPs should inquire about the impact of ADHD symptoms on household functioning, including task and chore completion, disorganization and clutter, and risky behaviors in the home. Additionally, family stress levels should be assessed as parents of children with ADHD often report higher levels of stress and feelings of inadequacy in managing their child's behaviors (Harpin 2005). These challenges can increase the risk for divorce, depression, and substance use (CADDRA 2020; Peasgood et al. n.d.). Siblings may also experience diminished quality of life due to the additional demands placed on the family system.

### 3.3.3 Social Functioning

Social impairments are prevalent across the lifespan in individuals with ADHD and have a profound impact on quality of life (Kosheleff et al. 2023). Difficulties commonly include forming and maintaining relationships, interpreting social cues, regulating emotional responses, and participating appropriately in social contexts.

Children with predominantly inattentive presentations of ADHD may struggle to maintain social relationships, often resulting in social withdrawal and diminished self-esteem (French et al. 2024). Over time, repeated experiences of social failure can contribute to the emergence of oppositional, aggressive, or even antisocial behaviors during adolescence and into adulthood (Harpin 2005). Given these risks, it is essential to assess for emotional sequelae—including internalizing symptoms, negative self-concept, and low self-esteem—as part of both diagnostic evaluation and ongoing monitoring of treatment response (CADDRA 2020). Comprehensive assessment should also include evaluation of peer relationships, friendship quality, social withdrawal, and any signs of antisocial behavior (Adamou et al. 2024; Kosheleff et al. 2023).

As individuals with ADHD enter adolescence and adulthood, distinct interpersonal challenges often become more pronounced. Compared to neurotypical peers, adults with ADHD are more likely to experience higher rates of divorce, remain unmarried, and report greater emotional loneliness (Michielsen et al. 2013). Impulsivity may show up in social contexts as inappropriate comments, reckless actions, or hastily agreed-upon commitments—behaviors that can erode trust in both personal and professional relationships (Morellini et al. 2022). Although many adults with ADHD retain the cognitive capacity to identify optimal solutions, they often struggle to generate effective responses during interpersonal conflicts, contributing to broader relational difficulties (Thoma et al. 2020).

Emotional dysregulation further complicates social interactions. Individuals may respond with overwhelming or inappropriate emotions, and rejection sensitivity can lead them to avoid social situations altogether (Morellini et al. 2022).

Daily functional impairments also affect communication and relational trust. Time blindness—marked by chronic lateness and missed obligations—can damage perceptions of reliability. Disorganization, such as clutter and forgetfulness, may reinforce stereotypes of irresponsibility. Inattentiveness during conversations—appearing distracted or forgetting shared details—can make others feel devalued (Neprily et al. 2025).

Over time, these cumulative difficulties can lead to social withdrawal, smaller support networks, and diminished quality of life. Social dysfunction is closely tied to comorbid anxiety and depression, amplifying impairment across developmental stages (Michielsen et al. 2013; Morellini et al. 2022).

### 3.3.4 Autonomy and Life Skills

Adolescents transitioning into young adulthood encounter a range of challenges that can impact their overall quality of life and functioning. This developmental stage is marked by significant shifts in academic, occupational, social, and healthcare environments. Adolescents are also forming deeper relationships, striving for greater independence, and developing a clearer sense of self. It is imperative that NPs consider these psychosocial changes when assessing adolescents and young adults. This period of development often includes increased risk-taking, greater sexual activity, and a shift in focus from family relationships toward peer (Boland et al. 2022).

Deficits in executive functions such as planning, organizing, and time management make the transition to independent living particularly difficult (Sibley 2023). Adolescents with ADHD may struggle more than their peers as they transition into adulthood due to a lack of structure and increased expectations (Solomon and Sparrow 2024). Of the 5–8% of children who are diagnosed with ADHD, roughly 60–85% continue to be symptomatic into adolescence, young adulthood, and adulthood (Caye et al. 2016; Howard et al. 2016). As such, assessment should include evaluation of life skills such as financial management, maintaining relationships, and independent self-care.

### 3.3.5 ADHD-Related Impairment in Older Adults

ADHD in older adults—typically referring to individuals in their late 50s, 60s, and beyond—is an emerging area of study. Although research remains limited, evidence suggests that older adults with ADHD continue to experience meaningful functional impairments, which may compound age-related challenges. Older adults with ADHD report poorer quality of life compared to their neurotypical peers (Dobrosavljevic et al. 2023). Functional impairments can include difficulties managing medical appointments, completing routine daily tasks, sustaining social relationships, and coping with retirement activities. Moreover, longstanding attention and memory challenges may exacerbate typical age-related cognitive decline. For NPs assessing older adults, it is crucial to differentiate impairments stemming from ADHD versus those due to aging or neurodegenerative conditions like dementia. In addition to the clinical interview, formal impairment measures should be considered to assist in this differentiation.

NPs can further enhance assessment accuracy by incorporating collateral information from spouses, adult children, or caregivers, as older adults may have limited insight into their functional challenges. It is important to evaluate daily living skills and establish a baseline for continuing to assess treatment outcomes. Experts recommend reviewing the patient's lifelong history, as many older adults with ADHD demonstrate an accumulated lifetime burden across education, occupational, and social domains (Brod et al. 2011).

## 3.4 Symptom Presentation and Associated Features of ADHD in Females

Despite growing recognition of ADHD in females, significant disparities remain in its identification, assessment, and treatment. Girls and women are frequently underdiagnosed or diagnosed later in life compared to males, often due to less overt symptom presentation, clinician bias, and the masking effects of comorbid internalizing disorders (Almekhlafi and Jain n.d.; Martin et al. 2023). These disparities are compounded by the "female protective effect," wherein girls must display more severe impairment before they are referred for assessment (Mowlem et al. 2019). As a result, many females receive alternative diagnoses, such as anxiety or depression, or are prescribed antidepressants prior to being accurately diagnosed with ADHD (Martin et al. 2023). Early and accurate identification is critical, as it can profoundly impact self-esteem, academic achievement, and long-term mental health.

Clinically, ADHD in females is characterized by distinct cognitive and behavioral patterns that frequently diverge from DSM-5 criteria. Common cognitive manifestations include persistent executive function difficulties (e.g., sustained attention, organization, and working memory deficits) and language processing impairments (Berry et al. 1985; Miller et al. 2013; Oroian et al. 2024). In addition, many females employ compensatory strategies such as perfectionism, overpreparation, and social mimicry and camouflaging that obscure their difficulties (Attoe and Climie 2023; Platania et al. n.d.; Quinn 2005). Females often appear "high functioning" externally, even when experiencing significant internal distress. Behavioral presentations are often dominated by internalizing symptoms—including anxiety, low self-esteem, social withdrawal, and emotional dysregulation—rather than the more conspicuous hyperactive-impulsive behaviors typically seen in males (Antoniou et al. 2021; Hinshaw et al. 2021). Adding further complexity, hormonal and developmental changes—such as those associated with puberty, menstrual cycles, pregnancy, and menopause—can significantly influence symptom severity and coping capacity in females with ADHD (Antoniou et al. 2021; Quinn 2005).

When evaluating females for ADHD, NPs should remain vigilant for chronic anxiety, depressive symptoms, or unexplained academic and occupational over- or under-performance. Screening for ADHD in women should be considered standard practice, much like screening for depression and anxiety, particularly when these internalizing symptoms dominate the clinical picture (Babinski and Libsack 2025). Brief screening tools such as the ASRS-v1.1 have demonstrated good sensitivity to the inattentive presentation commonly seen in females (Kessler et al. 2005; Ustun et al. 2017).

For comprehensive evaluation, use of gender-informed diagnostic tools is recommended. The DIVA-5 diagnostic interview explicitly probes gender differences in symptom expression, including emotional dysregulation and impairments in social functioning (Platania et al. n.d.; Young et al. 2020). The ADHD Rating Scale-IV, meanwhile, accounts for lower symptom thresholds in females while recognizing comparable levels of functional impairment (Rucklidge and Kaplan 1997).

In sum, accurate and timely ADHD assessment in females requires a nuanced, developmentally and gender-informed approach that goes beyond DSM-5-TR criteria. Incorporating knowledge of internalizing symptomatology, hormonal influences, compensatory behaviors, and social context is essential. NPs who adopt such an approach are well-positioned to improve diagnostic equity and deliver targeted interventions that can substantially alter the trajectory of ADHD in women and girls.

## 3.5 Recognizing Co-occurring Disorders Across the Lifespan

A critical part of a comprehensive ADHD evaluation is assessing for co-occurring conditions and conducting a careful differential diagnosis. Studies have shown that ADHD rarely occurs in isolation (Sonuga-Barke et al. 2023); in fact, up to 70–80% of children, adolescents, and adults with ADHD meet criteria for at least one additional psychiatric diagnosis (Danielson et al. 2024; Katzman et al. 2017).

Among children and adolescents, the most frequently observed co-occurring conditions are disruptive and impulse-control disorders, depressive disorders, and generalized anxiety, with some variation depending on ADHD subtype. Oppositional defiant disorder (ODD) and conduct disorder (CD) are particularly common, co-occurring with ADHD in approximately 30–50% of cases (Eskander n.d.), with hyperactive and combined subtypes at higher risk than inattentive presentations. Intermittent explosive disorder is also more common among individuals with ADHD compared to the general population (Kessler et al. 2006).

Other neurodevelopmental disorders frequently co-occur with ADHD. For example, an estimated 25–47% of children with ADHD also meet criteria for a learning disorder (Sonuga-Barke et al. 2023). Co-occurrence with autism spectrum disorder (ASD) varies widely, with reported rates ranging from 9% to 50% (CADDRA 2020; Yerys et al. 2009). Additionally, conditions such as Tourette's syndrome and other tic disorders are also commonly comorbid with ADHD, pointing to potentially shared neurobiological mechanisms (Huisman-van Dijk et al. 2016).

Adults with ADHD are at heightened risk for a range of psychiatric and medical comorbidities. Common co-occurring mental health conditions include anxiety, depression, self-harm, substance use, and increased utilization of emergency services (Langley et al. 2023). Approximately 18–33% meet criteria for depressive disorders, 19–35% for bipolar disorder (Choi et al. n.d.; Kessler et al. 2006), and up to 50% for an anxiety disorder (French et al. 2024). OCD also shows notable overlap, with 11–15% of adults and up to 26% of children and adolescents with OCD meeting criteria for ADHD (Cabarkapa et al. n.d.; Kushki et al. 2019). Substance use disorders (SUDs) are significantly more prevalent in this population, with about 20–25% of adults with SUD also having ADHD—a rate that tends to increase during adolescence and plateau in adulthood (Wilens et al. 2003). Sleep disturbances affect over half of individuals with ADHD and commonly present as delayed sleep onset, insomnia, hypersomnia, or sleep procrastination (Wynchank et al. 2018). Additionally, physical health issues such as obesity, epilepsy, autoimmune conditions, and allergies are reported at higher rates in those with ADHD compared to the general population (Leppert et al. 2021; Rietz et al. 2021).

## 3.6 Differential Diagnosis

Accurately ruling out other mental health and medical conditions is a critical component of a comprehensive ADHD evaluation. Differential diagnosis involves identifying overlapping symptoms and distinguishing ADHD from other conditions with similar presentations (see Table 3.5) (CADDRA 2020). Core ADHD symptoms are also common in mood and anxiety disorders, ASD, and other psychiatric conditions. Because there are no reliable biomarkers for mental health disorders, NPs must rely on a combination of subjective reports, objective measures, and clinical judgment, as outlined earlier in this chapter. While the DSM-5-TR requires

**Table 3.5** Differential diagnosis for ADHD

| Disorder | Overlapping features with ADHD | Differentiating features |
|---|---|---|
| Depression | Diminished ability to think/concentrate, poor motivation, indecisiveness | Typically later onset, episodic course |
| Mania | Distractibility, impulsivity, increased talkativeness, energy/restlessness | Later onset, episodic course, elated mood, rapid cycling, grandiosity, increased sexuality |
| Anxiety disorders | Distractibility, executive function difficulties, rumination, perfectionism, restlessness | Distraction occurs only in the presence of the anxiety focus (e.g., obsessional thinking in OCD, self-conscious ideation in social phobia); later onset than ADHD |
| Substance use disorders (SUDs) | Poor attention, distractibility, hyperactivity | Cognitive and behavioral impairments occur only in the context of substance use |
| Psychotic disorders (e.g., schizophrenia) | Prominent cognitive symptoms | Presence of hallucinations, delusions, and thought disorder |
| Oppositional defiant disorder (ODD) | Resistance to doing tasks at school or work | Defiance in response to authority figures |
| Intellectual developmental disorder (IDD) | Inattention, impulsivity, difficulty in social settings, and emotional dysregulation | Presence of cognitive delays |

(continued)

**Table 3.5** (continued)

| Disorder | Overlapping features with ADHD | Differentiating features |
|---|---|---|
| Autism spectrum disorder (ASD) | Inattention, fidgeting, impulsivity, and emotional dysregulation | Meltdowns/tantrums, problems with understanding emotions and relating to others, stereotypic movements (e.g., hand flapping) |
| Trauma disorders (i.e., reactive attachment disorder, PTSD) | Concentration difficulties, restlessness, inattention | Trauma response can cause dissociative states, making it difficult to concentrate |
| Personality disorders | Impulsivity, disorganization, social intrusiveness, emotional dysregulation, and cognitive dysregulation | Fear of abandonment, self-injury, extreme ambivalence |
| Neurocognitive disorders | Impairment in attention, memory, and other cognitive functions | Represents a decline from previous functioning/performance; typically adult onset |
| Medication induced symptoms (ex: bronchodilators, isoniazid, antipsychotics, thyroid replacement) | Inattention, hyperactivity, impulsiveness, restlessness | Symptoms occur after medication is started and subside when medication is reduced or discontinued |

medical conditions to be ruled out, this may occasionally warrant laboratory testing or a medical referral. Accurate diagnosis is essential for effective treatment planning, particularly when considering pharmacologic management. Referral for neuropsychological assessment, psychiatry, or additional medical evaluation may be necessary to clarify complex cases.

## 3.7 Considerations for Assessment of Diverse Populations and Underserved Communities

NPs play a critical role in addressing disparities in ADHD and mental health assessment by interpreting assessment data within a broader cultural and contextual framework. There are disparities in ADHD diagnosis across racial and ethnic groups. For instance, White children in the USA are more likely to be diagnosed with ADHD, whereas Hispanic and Black children are more frequently misdiagnosed with disruptive behavior disorders like ODD or CD—often due to provider interpretation shaped by a predominantly White cultural framework (Coker et al. 2016; Fadus et al. 2020; Shalaby et al. 2024; Shi et al. 2021).

Insurance status, cultural norms, and gender further influence diagnostic outcomes. White children with private insurance are most likely to receive diagnosis and treatment for ADHD, while Asian children show the lowest rates (Shi et al. 2021).

These disparities reflect gaps in access, provider bias, and systemic inequities. NPs must be trained to conduct equitable, trauma-informed assessments that consider these cultural and structural factors.

### 3.7.1 Differential Diagnosis and Cultural Context

Misdiagnosis between ADHD, ODD, and CD can reinforce systemic inequities, particularly among African American and Hispanic youth. While ADHD is characterized by inattention, hyperactivity, and impulsivity, ODD involves defiance and irritability, and CD includes aggressive or rights-violating behaviors (American Psychiatric Association 2022). These symptom clusters may appear similar but differ in origin and meaning. For example, behaviors associated with CD—such as perceiving others as hostile—may be adaptive responses to chronic exposure to violence or unsafe environments, rather than pathological (Fadus et al. 2020). Thus, culturally and contextually informed differential diagnosis is essential.

### 3.7.2 Addressing Bias and Enhancing Cultural Competence

Even clinicians with strong commitments to equity can be influenced by unconscious biases related to race, ethnicity, gender, sexuality, body size, ability, and more (Dovidio and Fiske 2012). These biases are often implicit and shaped by early social experiences and structural conditioning. Within healthcare, such biases have been shown to negatively affect diagnostic decisions, clinical interactions, and treatment recommendations (Dovidio and Fiske 2012; Fadus et al. 2020).

To mitigate these effects, NPs must remain vigilant and pursue ongoing training in cultural humility and structural competency. This includes recognizing how overlapping identities (e.g., race, gender, class, language status) create unique vulnerabilities to diagnostic bias and under-treatment. Tools such as the DSM-5 Cultural Formulation Interview (CFI) can support more accurate, culturally responsive assessments by helping clinicians explore the cultural, social, and environmental contexts of patients' symptoms (Fadus et al. 2020; Lewis-Fernández et al. 2014).

## 3.8 Final Remarks

In sum, the accurate assessment of ADHD across the lifespan requires a comprehensive, developmentally informed, and culturally responsive approach—one that goes far beyond symptom checklists. While US guidelines provide a strong foundation for pediatric ADHD, they fall short in addressing the complexity of adult presentations, co-occurring conditions, and systemic disparities. NPs are uniquely positioned to bridge these gaps due to their holistic training, accessibility, and role in multidisciplinary care teams. By integrating a multi-method, evidence-based clinical assessment, including diagnostic interviews, ADHD-specific rating scales, and functional impairment measures, NPs can enhance diagnostic accuracy and promote tailored, affirming interventions. Moreover, embracing neurodiversity, attending to gender- and culture-specific presentations, and rigorously screening for comorbidities allow providers to better serve the full spectrum of individuals with ADHD.

## References


Adamou M, Arif M, Asherson P, Cubbin S, Leaver L, Sedgwick-Müller J, Müller-Sedgwick U, van Rensburg K, Kustow J. The adult ADHD assessment quality assurance standard. Front Psychiatry. 2024;15:Article 1380410. https://doi.org/10.3389/fpsyt.2024.1380410.

Adler LA, Shaw DM. Diagnosing ADHD in adults. Atten Defic Hyperact Disord. 2011;3(1):1–10. https://doi.org/10.1007/s12402-010-0043-8.

Almekhlafi K, Jain S. Unveiling gender disparities in ADHD: a literature review on factors and impacts of late diagnosis in females (2010–2023)–Open Access Pub. n.d. Retrieved April 24, 2025, from https://openaccesspub.org/womans-mental-health/article/2125

American Psychiatric Association. Diagnostic and statistical manual of mental disorders: DSM-5-TR. 5th ed. Washington, DC: American Psychiatric Association Publishing; 2022.

Antoniou E, Rigas N, Orovou E, Papatrechas A, Sarella A. ADHD symptoms in females of childhood, adolescent, reproductive and menopause period. Materia Socio Medica. 2021;33(2):114. https://doi.org/10.5455/msm.2021.33.114-118.

Antshel KM, Zhang-James Y, Faraone SV. The comorbidity of ADHD and autism spectrum disorder. Taylor & Francis; 2013. https://www.tandfonline.com/doi/abs/10.1586/14737175.2013.840417

Attoe DE, Climie EA. Miss. Diagnosis: a systematic review of ADHD in adult women. J Atten Disord. 2023;27(7):645–57. https://doi.org/10.1177/10870547231161533.

Babinski DE, Libsack EJ. Adult diagnosis of ADHD in women: a mixed methods investigation. J Atten Disord. 2025;29(3):207–19. https://doi.org/10.1177/10870547241297897.

Barbaresi WJ, Campbell L, Diekroger EA, Froehlich TE, Liu YH, O'Malley E, Pelham WEJ, Power TJ, Zinner SH, Chan E. Society for Developmental and Behavioral Pediatrics Clinical Practice guideline for the assessment and treatment of children and adolescents with complex attention-deficit/hyperactivity disorder. J Dev Behav Pediatr. 2020;41(Suppl 2S):S35–57. https://doi.org/10.1097/DBP.0000000000000770.

Barkley RA. Executive functions: what they are, how they work, and why they evolved. New York: Guilford Press; 2011.

Barkley RA. ADHD and the nature of self-control. New York: Guilford Press; 2023.

Barkley RA, Fischer M. The role of emotional impulsiveness in adult ADHD: impairment and occupational functioning. J ADHD Relat Disord. 2010;17(5):407–19. https://doi.org/10.1177/1087054710365998.

Berry CA, Shaywitz SE, Shaywitz BA. Girls with attention deficit disorder: a silent minority? A report on behavioral and cognitive characteristics. Pediatrics. 1985;76(5):801–9. https://doi.org/10.1542/peds.76.5.801.

Boland RJ, Verduin ML, Ruiz P, Shah A. Kaplan & Sadock's synopsis of psychiatry (12th ed.). Philadelphia, PA: Wolters Kluwer; 2022.

Brod M, Schmitt E, Goodwin M, Hodgkins P, Niebler G. ADHD burden of illness in older adults: a life course perspective. Qual Life Res. 2011;21(5):Article 5. https://doi.org/10.1007/s11136-011-9981-9.

Brown TE. Brown Executive Function/Attention Scales (Brown EF/A Scales): Manual. San Antonio, TX: Pearson; 2019.

Burns GL, Becker SP. Sluggish cognitive tempo and ADHD symptoms in a nationally representative sample of U.S. children: differentiation using categorical and dimensional approaches. J Clin Child Adolesc Psychol. 2021;50(2):267–80. https://doi.org/10.1080/15374416.2019.1678165.

Cabarkapa S, King JA, Dowling N, Ng CH. Co-morbid obsessive–compulsive disorder and attention deficit hyperactivity disorder: neurobiological commonalities and treatment implications. Frontiers. n.d. https://doi.org/10.3389/fpsyt.2019.00557.

CADDRA. Canadian ADHD practice guidelines. 4.1 ed. Toronto: Canadian ADHD Resource Alliance; 2020.

Caroline S SS, Sudhir PM, Mehta UM, Kandasamy A, Thennarasu K, Benegal V. Assessing adult ADHD: an updated review of rating scales for adult attention deficit hyperactivity disorder (ADHD). J Atten Disord. 2024;28(7):1045–62. https://doi.org/10.1177/10870547241226654.

Caye A, Spadini AV, Karam RG, Grevet EH, Rovaris DL, Bau CHD, Rohde LA, Kieling C. Predictors of persistence of ADHD into adulthood: a systematic review of the literature and meta-analysis. Eur Child Adolesc Psychiatry. 2016;25(11):Article 11. https://doi.org/10.1007/s00787-016-0831-8.

Chamberlain SR, Derbyshire KL, Leppink EW, Grant JE. ADHD symptoms and self-esteem in young adults: a population-based study. J Atten Disord. 2016;20(10):934–43. https://doi.org/10.1177/1087054713498947.

Chan WWY, Shum KK, Sonuga-Barke EJS. Attention-deficit/hyperactivity disorder (ADHD) in cultural context: do parents in Hong Kong and the United Kingdom adopt different thresholds when rating symptoms, and if so why? Int J Methods Psychiatr Res. 2022;31(3):e1923. https://doi.org/10.1002/mpr.1923.

Choi W-S, Woo YS, Wang S-M, Lim HK, Bahk W-M. The prevalence of psychiatric comorbidities in adult ADHD compared with non-ADHD populations: a systematic literature review. n.d. https://doi.org/10.1371/journal.pone.0277175.

Coker TR, Elliott MN, Toomey SL, Schwebel DC, Cuccaro P, Tortolero Emery S, Davies SL, Visser SN, Schuster MA. Racial and Ethnic Disparities in ADHD Diagnosis and Treatment. Pediatrics. 2016;138(3). https://doi.org/10.1542/peds.2016-0407.

Danielson ML, Claussen AH, Bitsko RH, Katz SM, Newsome K, Blumberg SJ, Kogan MD, Ghandour R. ADHD prevalence among U.S. children and adolescents in 2022: diagnosis, severity, co-occurring disorders, and treatment. Taylor & Francis; 2024. https://www.tandfonline.com/doi/abs/10.1080/15374416.2024.2335625

DeSerisy M, Hirsch E, Roy AK. The contribution of sensory sensitivity to emotional lability in children with ADHD symptoms. Evid-Based Pract Child Adolesc Mental Health. 2019;4(4):319–27. https://doi.org/10.1080/23794925.2019.1647122.

Dobrosavljevic M, Larsson H, Cortese S. The diagnosis and treatment of attention-deficit hyperactivity disorder (ADHD) in older adults. Expert Rev Neurother. 2023;23(10):883–93. https://doi.org/10.1080/14737175.2023.2250913.

Dovidio JF, Fiske ST. Under the radar: how unexamined biases in decision-making processes in clinical interactions can contribute to health care disparities. Am J Public Health. 2012;102(5):945–52. https://doi.org/10.2105/AJPH.2011.300601.

Eom TH, Kim Y-H. Clinical practice guidelines for attention-deficit/hyperactivity disorder: recent updates. Clin Exp Pediatr. 2023;67(1):26–34. https://doi.org/10.3345/cep.2021.01466.

Eskander N. The psychosocial outcome of conduct and oppositional defiant disorder in children with attention deficit hyperactivity disorder. Cureus. n.d.;12(8):e9521. https://doi.org/10.7759/cureus.9521.

Fadus MC, Ginsburg KR, Sobowale K, Halliday-Boykins CA, Bryant BE, Gray KM, Squeglia LM. Unconscious bias and the diagnosis of disruptive behavior disorders and ADHD in African American and Hispanic youth. Acad Psychiatry. 2020;44(1):95–102. https://doi.org/10.1007/s40596-019-01127-6.

Faraone SV, Kuntsi J. Childhood predictors of adult ADHD: findings from the European Union-funded IMAGE project. Eur Child Adolesc Psychiatry. 2008;17(Suppl 1):I35–41. https://doi.org/10.1007/s00787-008-1005-7.

First MB, Williams JBW, Karg RS, Spitzer RL. Structured clinical interview for DSM-5 disorders–clinician version (SCID-5-CV). Arlington: American Psychiatric Publishing; 2015.

French B, Nalbant G, Wright H, Sayal K, Daley D, Groom MJ, Cassidy S, Hall CL. The impacts associated with having ADHD: an umbrella review. Front Psych. 2024;15:1343314. https://doi.org/10.3389/fpsyt.2024.1343314.

Fuermaier ABM, Gontijo-Santos Lima C, Tucha O. Impairment assessment in adult ADHD and related disorders: current opinions from clinic and research. J Atten Disord. 2024;28(12):1529–41. https://doi.org/10.1177/10870547241261598.

Gratz KL, Roemer L. Multidimensional assessment of emotion regulation and dysregulation: development, factor structure, and initial validation of the difficulties in emotion regulation scale. J Psychopathol Behav Assess. 2004;26(1):Article 1.

Grossman A, Avital A. Emotional and sensory dysregulation as a possible missing link in attention deficit hyperactivity disorder: a review. Front Behav Neurosci. 2023;17:1118937. https://doi.org/10.3389/fnbeh.2023.1118937.

Harpin V. The effect of ADHD on the life of an individual, their family, and community from preschool to adult life. Arch Dis Child. 2005;90(Suppl 1):i2–7. https://doi.org/10.1136/adc.2004.059006.

Herbert JD, Forman EM, England EL. Psychological assessment. In: Hersen M, Beidel D, editors. Adult psychopathology and diagnosis. Hoboken: Wiley; 2010. p. 91–134.

Hinshaw SP, Nguyen PT, O'Grady SM, Rosenthal EA. Annual research review: attention-deficit/hyperactivity disorder in girls and women: underrepresentation, longitudinal processes, and key directions. J Child Psychol Psychiatry. 2021;63(4):484–96. https://doi.org/10.1111/jcpp.13480.

Howard AL, Strickland N, Murray DW, Tamm L, Swanson JM, Hinshaw SP, Arnold LE, Molina BSG. Progression of impairment in adolescents with ADHD through the transition out of high school: contributions of parent involvement and college attendance. J Abnorm Psychol. 2016;125(2):233–47. https://doi.org/10.1037/abn0000100.

Huisman-van Dijk HM, van de Schoot R, Rijkeboer MM, Mathews CA, Cath DC. The relationship between tics, OC, ADHD and autism symptoms: a cross-disorder symptom analysis in Gilles de la Tourette syndrome patients and their family members. Psychiatry Res. 2016;237:138–46. https://doi.org/10.1016/j.psychres.2016.01.051.

Jangmo A, Stålhandske A, Chang Z, Chen Q, Almqvist C, Feldman I, Bulik CM, Lichtenstein P, D'Onofrio B, Kuja-Halkola R, Larsson H. Attention-deficit/hyperactivity disorder, school performance and effect of medication. J Am Acad Child Adolesc Psychiatry. 2019;58(4):423–32. https://doi.org/10.1016/j.jaac.2018.11.014.

Kapp SK, Gillespie-Lynch K, Sherman LE, Hutman T. Deficit, difference, or both? Autism and neurodiversity. Dev Psychol. 2013;49(1):59–71. https://doi.org/10.1037/a0028353.

Katzman MA, Bilkey TS, Chokka PR, Fallu A, Klassen LJ. Adult ADHD and comorbid disorders: clinical implications of a dimensional approach. BMC Psychiatry. 2017;17(1):302. https://doi.org/10.1186/s12888-017-1463-3.

Kaufman J, Birmaher B, Brent D, Rao U, Flynn C, Moreci P, Williamson D, Ryan N. Schedule for affective disorders and schizophrenia for school-age children-present and lifetime version (K-SADS-PL): initial reliability and validity data. J Am Acad Child Adolesc Psychiatry. 1997;36(7):980–8. https://doi.org/10.1097/00004583-199707000-00021.

Kennedy RJ, Quinlan DM, Brown TE. Comparison of two measures of working memory impairments in 220 adolescents and adults with ADHD. Journal of Attention Disorders. 2019:23(14):1793–1801. https://doi.org/10.1177/1087054716661232.

Kessler RC, Adler L, Ames M, Demler O, Faraone S, Hiripi E, Howes MJ, Jin R, Secnik K, Spencer T, Ustun TB, Walters EE. The World Health Organization adult ADHD self-report scale (ASRS): a short screening scale for use in the general population. Psychol Med. 2005;35(2):2245–56. https://doi.org/10.1017/S0033291704002892. Cambridge Core

Kessler RC, Adler L, Barkley R, Biederman J, Conners CK, Demler O, Faraone SV, Greenhill LL, Howes MJ, Secnik K, Spencer T, Ustun TB, Walters EE, Zaslavsky AM. The prevalence and correlates of adult ADHD in the United States: results from the National Comorbidity Survey Replication. Am J Psychiatry. 2006;163(4):716–23. https://doi.org/10.1176/ajp.2006.163.4.716.

Kooij JJS, Bijlenga D, Salerno L, Jaeschke R, Bitter I, Asherson P. Updated European Consensus Statement on diagnosis and treatment of adult ADHD. Eur Psychiatry. 2019;56:14–34.

Kosheleff AR, Mason O, Jain R, Koch J, Rubin J. Functional impairments associated with ADHD in adulthood and the impact of pharmacological treatment. 2023. https://pmc.ncbi.nlm.nih.gov/articles/PMC10173356/

Kushki A, Anagnostou E, Hammill C, Duez P, Brian J, Iaboni A, Schachar R, Crosbie J, Arnold P, Lerch JP. Examining overlap and homogeneity in ASD, ADHD, and OCD: a data-driven, diagnosis-agnostic approach. Transl Psychiatry. 2019;9(1):318. https://doi.org/10.1038/s41398-019-0631-2.

Langley K, Pozo-Banos MD, Daalsgard S, Paranjothy S, Riglin L, John A, Thapar A. ADHD and autism spectrum disorder (ASD) in childhood: establishing the feasibility and validity of a nation-wide e-cohort. medRxiv, 2023.01.10.23284395. 2023. https://doi.org/10.1101/2023.01.10.23284395.

Leppert B, Riglin L, Wootton RE, Dardani C, Thapar A, Staley JR, Tilling K, Davey Smith G, Thapar A, Stergiakouli E. The effect of attention deficit/hyperactivity disorder on physical health outcomes: A 2-sample Mendelian randomization study. Am J Epidemiol. 2021;190(6):1047–55. https://doi.org/10.1093/aje/kwaa273.

Lewis-Fernández R, Aggarwal NK, Bäärnhielm S, Rohlof H, Kirmayer LJ, Weiss MG, Jadhav S, Hinton L, Alarcón RD, Bhugra D, Groen S, van Dijk R, Qureshi A, Collazos F, Rousseau C, Caballero L, Ramos M, Lu F. Culture and psychiatric evaluation: operationalizing cultural formulation for DSM-5. Psychiatry. 2014;77(2):130–54. https://doi.org/10.1521/psyc.2014.77.2.130.

Martel MM, Eng AG, Bansal PS, Smith TE, Elkins AR, Goh PK. Multiple informant average integration of ADHD symptom ratings predictive of concurrent and longitudinal impairment. Psychol Assess. 2021;33(5):443–51. https://doi.org/10.1037/pas0000994.

Martin J, Langley K, Cooper M, Rouquette OY, John A, Sayal K, Ford T, Thapar A. Sex differences in ADHD diagnosis and clinical care: a national study of population healthcare records in Wales. 2023. https://doi.org/10.1101/2023.10.20.23297324.

Michielsen M, Comijs HC, Semeijn EJ, Beekman ATF, Deeg DJH, Sandra Kooij JJ. The comorbidity of anxiety and depressive symptoms in older adults with attention-deficit/hyperactivity disorder: a longitudinal study. J Affect Disord. 2013;148(2):220–7. https://doi.org/10.1016/j.jad.2012.11.063.

Miller M, Loya F, Hinshaw SP. Executive functions in girls with and without childhood ADHD: developmental trajectories and associations with symptom change. J Child Psychol Psychiatry. 2013;54(9):1005–15. https://doi.org/10.1111/jcpp.12074.

Morellini L, Ceroni M, Rossi S, Zerboni G, Rege-Colet L, Biglia E, Morese R, Sacco L. Social cognition in adult ADHD: a systematic review. Frontiers. 2022; https://doi.org/10.3389/fpsyg.2022.940445.

Mowlem F, Agnew-Blais J, Taylor E, Asherson P. Do different factors influence whether girls versus boys meet ADHD diagnostic criteria? Sex differences among children with high ADHD symptoms. Psychiatry Res. 2019;272:765–73. https://doi.org/10.1016/j.psychres.2018.12.128.

Mueller M, Segal DL. Structured versus unstructured diagnostic interviews: a review of the literature. Clin Psychol Sci Pract. 2014;21(2):177–91.

Müller V, Mellor D, Pikó BF. Associations between ADHD symptoms and rejection sensitivity in college students: exploring a path model with indicators of mental well-being. Learn Disabil Res Pract. 2024;39(4):223–36. https://doi.org/10.1177/09388982241271511.

Mulraney M, Arrondo G, Musullulu H, Iturmendi-Sabater I, Cortese S, Westwood SJ, Donno F, Banaschewski T, Simonoff E, Zuddas A, Döpfner M, Hinshaw SP, Coghill D. Systematic review and meta-analysis: screening tools for attention-deficit/hyperactivity disorder in children and adolescents. n.d. Retrieved May 1, 2025, from https://www.jaacap.org/article/S0890-8567(21)02084-0/abstract

Neprily KM, Climie EA, McCrimmon A, Makarenko E. Why can't we be friends? A narrative review of the challenges of making and keeping friends for children and adolescents with attention-deficit/hyperactivity disorder. Frontiers. 2025; https://doi.org/10.3389/fdpys.2024.1390791.

Nigg JT, Willcutt EG, Doyle AE, Sonuga-Barke EJS. Causal heterogeneity in attention-deficit/hyperactivity disorder: do we need neuropsychologically impaired subtypes? Biol Psychiatry. 2005;57(11):1224–30. https://doi.org/10.1016/j.biopsych.2004.08.025.

Oroian BA, Costandache G, Popescu E, Nechita P, Szalontay A. The uncharted territory of female adult ADHD: a comprehensive review. Eur Psychiatry. 2024;67(S1):S299–300. https://doi.org/10.1192/j.eurpsy.2024.624.

Panagiotidi M, Overton P, Stafford T. Sensory sensitivity as a predictor of ADHD traits in the general population. J Atten Disord. 2018;22(8):734–43. https://doi.org/10.1177/1087054716664409.

Peasgood T, Bhardwaj A, Biggs K, Brazier JE, Coghill D, Cooper CL, Daley D, Silva CD, Harpin V, Hodgkins P, Nadkarni A, Setyawan J, Sonuga-Barke EJS. The impact of ADHD on the health and well-being of ADHD children and their siblings. n.d. Retrieved May 1, 2025, from https://pmc.ncbi.nlm.nih.gov/articles/PMC5083759/

Pfeiffer B, Daly B, Nicholls L, Gullo MJ. Sensory processing patterns in children with ADHD and their correlation with behavioral symptoms. Occup Ther Int. 2015;22(1):25–35. https://doi.org/10.1002/oti.v22n1p25.

Platania NM, Starreveld DEJ, Wynchank D, Beekman ATF, Kooij S. Bias by gender: exploring gender-based differences in the endorsement of ADHD symptoms and impairment among adult patients. Frontiers. n.d. https://doi.org/10.3389/fgwh.2025.1549028.

Quinn PO. Treating adolescent girls and women with ADHD: gender-specific issues. J Clin Psychiatry. 2005;61(5):579–87. https://doi.org/10.1002/jclp.20121.

Rietz ED, Brikell I, Butwicka A, Leone M, Chang Z, Cortese S, D'Onofrio BM, Hartman CA, Lichtenstein P, Faraone SV, Kuja-Halkola R, Larsson H. Mapping phenotypic and aetiological associations between ADHD and physical conditions in adulthood in Sweden: a genetically informed register study. Lancet Psychiatry. 2021;8(9):774–83.

Robertson IH, Manly T, Andrade J, Baddeley BT, Yiend J. 'Oops!': performance correlates of everyday attentional failures in traumatic brain injured and normal subjects. Neuropsychologia. 1997;35(6):747–58. https://doi.org/10.1016/S0028-3932(97)00015-8.

Rucklidge JJ, Kaplan BJ. Psychological functioning of women identified in adulthood with attention-deficit/hyperactivity disorder. J Atten Disord. 1997;2(3):167–76. https://doi.org/10.1177/108705479700200303.

Sadozai AK, Sun C, Demetriou EA, Lampit A, Munro M, Perry N, Boulton KA, Guastella AJ. Executive function in children with neurodevelopmental conditions: a systematic review and meta-analysis. Nat Hum Behav. 2024;8(12):2357–66. https://doi.org/10.1038/s41562-024-02000-9.

Sasser T, Schoenfelder EN, Stein MA. Targeting functional impairments in the treatment of children and adolescents with ADHD. CNS Drugs. 2016;31(2):Article 2. https://doi.org/10.1007/s40263-016-0400-1.

Shalaby N, Sengupta S, Williams JB. Large-scale analysis reveals racial disparities in the prevalence of ADHD and conduct disorders. Sci Rep. 2024;14(1):25123. https://doi.org/10.1038/s41598-024-75954-5.

Shaw P, Stringaris A, Nigg J, Leibenluft E. Emotion dysregulation in attention deficit hyperactivity disorder. Am J Psychiatry. 2014;171(3):276–93. https://doi.org/10.1176/appi.ajp.2013.13070966.

Shi Y, Hunter Guevara LR, Dykhoff HJ, Sangaralingham LR, Phelan S, Zaccariello MJ, Warner DO. Racial disparities in diagnosis of attention-deficit/hyperactivity disorder in a US National Birth Cohort. JAMA Netw Open. 2021;4(3):e210321. https://doi.org/10.1001/jamanetworkopen.2021.0321.

Sibley M. Launching successful young adults with ADHD. CHADD; 2023. https://chadd.org/attention-article/launching-successful-young-adults-with-adhd/

Sibley MH, Arnold LE, Swanson JM, Hechtman LT, Kennedy TM, Owens E, Molina BSG, Jensen PS, Hinshaw SP, Roy A, Chronis-Tuscano A, Newcorn JH, Rohde LA. Variable patterns of remission from ADHD in the multimodal treatment study of ADHD. Am J Psychiatry. 2022;179(2):142–51. https://doi.org/10.1176/appi.ajp.2021.21010032.

Silverstein MJ, Faraone SV, Alperin S. How informant discrepancies can affect clinical assessment of ADHD. J Atten Disord. 2020;24(7):933–43.

Society for Developmental and Behavioral Pediatrics. Guidelines for the evaluation and management of ADHD in children with complex presentations. J Dev Behav Pediatr. 2020;41(3):145–61. https://doi.org/10.1097/DBP.0000000000000768.

Soler-Gutiérrez A-M, Pérez-González J-C, Mayas J. Evidence of emotion dysregulation as a core symptom of adult ADHD: a systematic review. n.d. https://doi.org/10.1371/journal.pone.0280131.

Solomon DEP, Sparrow EP. ADHD transitions: navigating from adolescence into early adulthood. Multi-Health Systems (MHS); 2024. https://mhs.com/blog/adhd-transitions-navigating-from-adolescence-into-early-adulthood/

Sonuga-Barke EJS, Becker SP, Bölte S, Castellanos FX, Franke B, Newcorn JH, Nigg JT, Rohde LA, Simonoff E. Annual research review: perspectives on progress in ADHD science—from characterization to cause. J Child Psychol Psychiatry. 2023;64(4):506–32. https://doi.org/10.1111/jcpp.13696.

Stibbe T, Huang J, Paucke M, Ulke C, Strauss M. Gender differences in adult ADHD: cognitive function assessed by the test of attentional performance. PLoS One. 2020;15(10):e0240810. https://doi.org/10.1371/journal.pone.0240810.

Surman C, Hammerness P, Faraone SV. Rejection sensitivity and emotional dysregulation in ADHD: A clinical perspective. J Clin Psychiatry. 2013;74(8):715–22. https://doi.org/10.4088/JCP.13m08596.

Szuromi B, Bitter I, Czobor P. Functional impairment in adult attention-deficit/hyperactivity disorder: the role of self-esteem and self-concept. J Atten Disord. 2013;17(5):333–41. https://doi.org/10.1177/1087054711436004.

Thoma P, Sonnenburg S, Marcinkowski N, Juckel G, Edel M-A, Suchan B. Social problem solving in adult patients with attention deficit hyperactivity disorder. Psychiatry Res. 2020;285:112721. https://doi.org/10.1016/j.psychres.2019.112721.

Toplak ME, West RF, Stanovich KE. Practitioner review: do performance-based measures and ratings of executive function assess the same construct? J Child Psychol Psychiatry. 2013;54(2):131–43. https://doi.org/10.1111/jcpp.12001.

Trivedi H, Stiebel V. The role of social determinants of health in the assessment and diagnosis of ADHD. Curr Psychiatry Rep. 2020;22(11):64.

Ustun B, Adler LA, Rudin C, Faraone SV, Spencer TJ, Berglund P, Kessler RC. The World Health Organization Adult ADHD Self-Report Scale (ASRS): a short screening scale for use in the general population. Psychol Med. 2017;47(2):1–10.

Weiss MD. A paradigm for targeting functional impairment as an outcome in attention-deficit/hyperactivity disorder. Brain Sci. 2022;12(8):1014. https://doi.org/10.3390/brainsci12081014.

Weiss MD, McBride NM, Craig S, Jensen P. Conceptual review of measuring functional impairment: findings from the Weiss Functional Impairment Rating Scale. Evid Based Ment Health. 2018;21(4):155–64. https://doi.org/10.1136/ebmental-2018-300025.

Wilens TE, Biederman J, Forkner P, Ditterline J, Morris M, Moore H, Galdo M, Spencer TJ, Wozniak J. Patterns of comorbidity and dysfunction in clinically referred preschool and school-age children with bipolar disorder. J Child Adolesc Psychopharmacol. 2003;13(4):495–505. https://doi.org/10.1089/104454603322724887.

Willcutt EG. The prevalence of DSM-IV attention-deficit/hyperactivity disorder: A meta-analytic review. Neurotherapeutics. 2012;9(3):490–502.

Wolraich ML, Hagan JF, Allan C, Chan E, Davison D, Earls M, Zurhellen W. Clinical practice guideline for the diagnosis, evaluation, and treatment of attention-deficit/hyperactivity disorder in children and adolescents. Pediatrics. 2019;144(4):e20192528. https://doi.org/10.1542/peds.2019-2528.

Wynchank D, ten HM, Bijlenga D, Penninx BW, Beekman AT, Lamers F, de GR, Kooij JJS. The association between insomnia and sleep duration in adults with attention-deficit hyperactivity disorder: results from a general population study. J Clin Sleep Med. 2018;14(03):349–57. https://doi.org/10.5664/jcsm.6976.

Yerys BE, Wallace GL, Sokoloff JL, Shook DA, James JD, Kenworthy L. Attention deficit/hyperactivity disorder symptoms moderate cognition and behavior in children with autism spectrum disorders. Autism Res. 2009;2(6):322–33. https://doi.org/10.1002/aur.103.

Young S, Adamo N, Ásgeirsdóttir BB, Branney P, Beckett M, Colley W, Cubbin S, Deeley Q, Farrag E, Gudjonsson G, Hill P, Hollingdale J, Kilic O, Lloyd T, Mason P, Paliokosta E, Perecherla S, Sedgwick J, Skirrow C, et al. Females with ADHD: an expert consensus statement taking a lifespan approach providing guidance for the identification and treatment of attention-deficit/hyperactivity disorder in girls and women. BMC Psychiatry. 2020;20(1):404. https://doi.org/10.1186/s12888-020-02707-9.

# 4 Establishing a Collaborative Care Model for ADHD

Nicholas M. Cutro, Jean Tidd, and Kathleen T. McCoy

## 4.1 Integration of Primary Care and Behavioral Health Teams

ADHD is one of the two most common neurodevelopmental disorders impacting school-age children (Liu et al. 2024). ADHD often co-occurs with other psychiatric conditions and neurodevelopmental disorders such as learning disabilities, anxiety disorders, oppositional defiant disorder, mood disorders, and substance use disorders. Comorbidity is the norm and not the exception. As a result, the diagnosis and treatment of ADHD is often complex, highlighting the importance of a comprehensive evaluation when evaluating an individual for suspected ADHD.

The benefits of integrated care cannot be overstated. Integrated care can lead to comprehensive assessment, accurate diagnosis, coordinated treatment planning and implementation, improved communication between providers and families, and enhanced treatment adherence and outcomes. Primary care providers (PCPs) may often be the first line of screening for ADHD and related disorders. This is an important point because identification and early intervention are essential when treating neurodevelopmental disorders such as ADHD.

N. M. Cutro (✉)
Attention, Intellectual & Learning Evaluations, LLC, Downtown–Orlando, FL, USA
e-mail: Testing@NickCutro.org

J. Tidd
Holistic Consultation LLC, Columbus, OH, USA
e-mail: jean@badassadhdcoaching.com

K. T. McCoy
Community Mental Health College of Nursing, Health Sciences (HAHN), University of South Alabama, Mobile, AL, USA
e-mail: mccoy@southalabama.edu

© The Author(s), under exclusive license to Springer Nature Switzerland AG 2026
R. J. Kennedy (ed.), *Assessment & Treatment of ADHD*,
https://doi.org/10.1007/978-3-032-24171-9_4

The roles and responsibilities of primary care teams can serve as the first line of defense and intervention. Initial screening for ADHD and related disorders can have a huge impact on identification. This can lead to targeted referrals to medical and behavioral health professionals. Behavioral health professionals may include school psychologists, pediatric neuropsychologists, social workers, and therapists. When ADHD and related disorders are flagged during a PCP visit, this can lead to targeted referrals to school teams, psychiatrists, and neuropsychologists who can further evaluate the whole individual to put forth a comprehensive treatment approach.

ADHD significantly affects multiple aspects of life, including academic achievement, occupational success, emotional regulation, and interpersonal relationships (Barkley 2015). Its understanding has evolved from a narrowly defined medical condition to a complex disorder with familial, social, and educational implications. Research increasingly highlights the increased health risks associated with ADHD, including a significantly reduced life expectancy. Studies show that children with ADHD are twice as likely to die in childhood, while adults with the disorder face a three- to fivefold increase in premature mortality by midlife. The leading cause of early death among individuals with ADHD is accidental injury, underscoring the urgent need for increased awareness, early intervention, and comprehensive support.

Stimulant medications are still one of the most effective treatments for managing ADHD symptoms, helping to regulate dopamine and norepinephrine levels in the brain to improve attention, impulse control, and executive functioning (Faraone et al. 2021). Despite their effectiveness, adherence to ADHD medication is a challenge due to biological, psychological, social, and systemic barriers. Many individuals, particularly adolescents and adults, struggle with maintaining consistent treatment, often due to side effects, stigma, misconceptions about medication, or executive functioning difficulties (Matza et al. 2022). While medication is not a cure for ADHD, it is a critical part of symptom management, enabling children and adults to navigate daily life more effectively. Without proper treatment, individuals face an increased risk of academic struggles, impulsive decision-making, emotional instability, and engagement in risky behaviors. Therefore, improving treatment adherence is essential for decreasing these negative outcomes and enhancing overall well-being.

Integrating primary care and behavioral health services improves the diagnosis, treatment, and long-term management of ADHD. PCPs are often the first point of contact for children and adults with ADHD, making them instrumental in early screening and diagnosis. Effective ADHD management typically needs a multifaceted approach, incorporating medication, behavioral therapy, psychoeducation, and lifestyle modifications. This collaboration ensures that individuals receive personalized, well-coordinated care that maximizes treatment effectiveness, improves adherence, and enhances overall quality of life.

By fostering a holistic and interdisciplinary approach, collaborative care not only improves ADHD outcomes but also reduces long-term risks associated with the disorder. Increased integration between medical and behavioral health services is essential for bridging gaps in treatment, reducing stigma, and ensuring that individuals with ADHD receive the comprehensive support they need to thrive.

## 4.2 Exploring Philosophical Underpinnings, Nursing Guidelines, and Appropriate Models

Attention deficit hyperactivity disorder (ADHD) is a neurodevelopmental condition that affects both children and adults, characterized by persistent patterns of inattention, hyperactivity, and impulsivity (APA 2022, pp. 68–70). The management of ADHD requires a multifaceted and collaborative approach to ensure effective care and support for individuals of any age and differ individually. It takes “a village” to assist children and adults by inference (Wolraich 2016). This “village” includes and is highly dependent upon contextual elements of the life of each affected individual, family, and major academic, work, and social and healthcare-associated environments.

### 4.2.1 Philosophical Underpinnings

The philosophy of care for individuals with ADHD is grounded in holistic and patient-centered principles. This approach recognizes the inherent complexity of ADHD and need for comprehensive care addressing physical, emotional, social, and psychological aspects of the condition (Wolraich 2016). The following philosophical concepts are foundational and crucial to ADHD care.

### 4.2.2 Holism

Holism emphasizes the importance of treating individuals, as a whole person, rather than focusing solely on symptoms of ADHD. This approach involves understanding the interconnectedness of the mind, body, and environment and how these aspects influence one another and individual functioning levels. Holism considers factors of an exhaustive span, from medical to societal influences (Weissenberger et al. 2017), each one potential to be a pivotal part of care.

### 4.2.3 Patient-Centered Care

Patient-centered care prioritizes the needs, preferences, and values of the individual with ADHD. It involves active, forthright communication and collaboration between the patient, the associated family/network, and healthcare providers to create a tailored care plan respecting unique experiences and goals of patients and families and the contextual elements of culture and faith/belief system. Patient-centered care ideally begins at initial inquiry related to symptoms and the scheduled diagnostic evaluative process. For initial visits for diagnosis and treatment with providers, identified patients and families are encouraged to communicate honestly with providers, especially the primary care provider. An organized prepared for the workup, ahead of the encounter, with a list of questions and a trusted family member or friend to assist with information may be very helpful. Speaking to concerns raised before, during, and after the encounter is important, because this way, the patient will have a voice in the care plan. Schatz et al., in the Systematic Review of Patients' and Parents' Preferences for ADHD Treatment Options and Processes of Care (2015), found that while more research on patient preference is warranted, patient preference, while dynamic, may influence better relationship with providers and care team members, enhancing fuller participation in care, therefore improving outcomes when the plan is established.

### 4.2.4 Interdisciplinary Collaboration and Coordination of Care

Interdisciplinary collaboration is essential in managing ADHD, as it involves the integration of various healthcare professionals, including but not limited to physicians, advance practice providers, nurses, psychologists, educators, and social workers. A collaborative effort better ensures that all aspects of patient care and involved family system are addressed comprehensively. However, this model must also the circle of influence, of the patient and family. Cropley and Duis Sanders discuss this model, which is not new to nursing. Nursing care has historically been direct, and that of coordination of care, particularly in inpatient and outpatient settings. The role of the Nurse Practitioner is that of communicator with all parties concerned with the patient, the plan of care, and outcomes. The Texas Nurses Association Coordination of Care Model (2013) illustrates the overarching stretch, the extended reach that the Coordination of Care Model will encompass to foster collaboration. These areas of care include but are not limited to the following: patient, primary care, healthcare teams, individual providers, specialty care, community care, academic/school districts, worship community, sports activities, environment, and "other," as appropriate to care. The role requires expert cognitive decision-making, communication and counseling skills, to adequately address the needs of the patient throughout the care continuum (Cropley et al. 2013).

## 4.3 Established Time-Honored Nursing Guidelines

Nursing guidelines for the care of individuals with ADHD have been developed and refined over the years to provide structured and effective support. Some key guidelines include the following:

### 4.3.1 Assessment and Diagnosis

The Nurse Practitioner plays a critical role in the initial assessment and diagnosis of ADHD. This involves gathering comprehensive patient histories, conducting behavioral assessments, and collaborating with other healthcare providers to ensure an accurate diagnosis. Scope of Practice may differ if the Nurse Practitioner is prepared as an Advanced Practice Nurse (APN), as Clinical Nurse Specialists and Nurse Practitioners are population-based. To clarify, the Psychiatric Mental Health Nurse Practitioner and Family Nurse Practitioner roles include across the lifespan. Adult and Geriatric Nurse Practitioners and Pediatric Nurse Practitioners are limited to defining ages of those populations. Each has a place in treating those with ADHD.

### 4.3.2 Education and Advocacy

Nurses provide education to patients and families about ADHD, including symptoms, treatment options, and coping strategies. Nurses anticipate and advocate for patient needs within the healthcare system and broader community.

### 4.3.3 Medication Management and Treatment Plan

Nurses participate on several levels and are responsible for working with the entire care team. To truly advocate for better patient outcomes, the nurse, by virtue of being central in the treatment, is called, by the 2021 4th Edition of the Scope and Standards of Nursing Practice (ANA), to work well with all team members. This, at times, requires tact, diplomacy, and ambassadorship. The Nurse Practitioner monitors treatment related to the entire treatment plan, including but not limited to medication, managing side effects, and educating patients and families about the appropriate use of compounds prescribed for ADHD and encourage full participation in all aspects of the plan of care. The APN also diagnoses and makes holistic treatment recommendations in partnership with the patient, family, and other healthcare team members. In addition to prescribing medications if needed, the APN provides selected therapies and makes referrals, all dependent upon the presiding local State Nurse Practice Act (NCSBN n.d.).

### 4.3.4 The Domain of Interprofessional Partnerships

The American Association of Colleges of Nursing (AACN n.d.) portrays collegial relationships in the context of care to be pivotal and has provided domain mastery for education of nurses on both entry and advance practice levels to promote intentional collaboration across interprofessionals, intraprofessionals, and paraprofessionals, the care team, patients, family, community, and other stakeholders to optimize care and facilitate healthcare and outcomes. Effective nurse behaviors effectively orchestrate the many elements of effective collaboration manifesting core professional values of altruism, excellence, caring, ethics, respect, communication, and shared accountability.

For the Advanced Practice Nurse, the American Association of Colleges of Nursing holds to the aforementioned for Nurse Practitioner practice and additional essentials in this area of care with foundational knowledge encompassing knowledge and scope of practice of collaborating professionals. These skills include the following: communication and collaborative skills in partnership, conflict resolution, supervision and oversight of practice, navigation of site-specific variations of care across the care continuum, an understanding and mastery of team dynamics and fluid leadership based upon care contexts, and knowledge of the Interprofessional Education competencies and frameworks.

### 4.3.5 Behavioral Interventions

Nurses implement behavioral interventions to support patients in managing ADHD symptoms. This includes techniques and processes such as positive reinforcement, structured routines, brief therapies, cognitive-behavioral therapies and strategies, and implementation of motivational interviewing.

### 4.3.6 Support and Counseling

Nurses offer emotional support and counseling to individuals with ADHD and families, providing safe space for patients and families to discuss both challenges and successes, offering guidance on coping mechanisms, stress management, resources, and referrals.

## 4.4 Models of Care

As previously mentioned in the section Interdisciplinary Collaboration and Coordination of Care, every effort is made to make care as seamless as possible, with consideration to synthesizing potential for many care contributors. Several models have been designed and developed to facilitate such care, each having its advantages; these are as follows:

### 4.4.1 Medical Home Model

The Medical Home Model is a primary care model, a patient-centered approach providing continuous, comprehensive care for individuals with ADHD involving a team of healthcare providers working collaboratively to coordinate care, manage chronic conditions, and ensure that all aspects of patient health are addressed. Ronis et al. (2015) discussed the burden of ADHD on families and the positive aspects of this style of care delivery.

### 4.4.2 Integrated Care Model

The Integrated Care Model focuses upon integration of primary care and mental health services ensuring individuals with ADHD receive holistic care addressing both physical and mental health needs.

### 4.4.3 Collaborative Care Model

The Collaborative Care Model involves the coordination of care between primary care providers, mental health specialists, and other healthcare professionals, emphasizing regular communication and collaboration to ensure that patient care is orchestrated as well-coordinated and comprehensive.

The mentioned models are discussed at length, in addition to three other models by Shahidullah et al. in a systematic review related to coordination of care for children with ADHD. Pediatric models have been studied for effectiveness and compared/contrasted. There are fewer models of Adult ADHD Primary Care Models; however, Casey et al. (2020) make a good case as pharmacists for adult models of coordinated care. Clearly, more research is needed for both children and adults.

## 4.5 School-Based Interventions

For children with ADHD, school-based interventions are crucial. These interventions involve collaboration between educators, school nurses, psychologists, and parents to create an inclusive and supportive learning environment that accommodates the needs of students with ADHD. This may include collaboration with a federally funded school district to create and establish an individualized education plan (MacLeod et al. 2017). Young adults transitioning into the workplace, higher education, or adult care may need collaborative approaches for optimal outcomes per Lasky et al. (2016).

### 4.5.1 Family-Centered Care Model

The Family-Centered Care Model recognizes the important role of the family in managing ADHD. This model involves actively engaging the family as central in the care process, providing education and support, and encouraging family members to participate in decision-making and care planning per Davis et al. (2012). Children with ADHD have special needs that will include the entire family, likewise, parents, adults, and adult workers with ADHD continue to have needs which may include the entire family and employers and, if appropriate, other members within the patient sphere of influence. In a scoping review of this model, by Kokorelias et al. (2019), it was determined that across all family-centered care models, the consistent goal is to develop and implement patient care plans within family contexts and that most aspects are universal, while many are condition/illness specific. While more research is needed, key components to facilitate family-centered care currently include (1) collaboration between family members and healthcare providers, (2) consideration of family contexts, (3) agency and jurisdictional policies and procedures, and (4) patient, family, and healthcare professional education.

## 4.6 Perspectives from Various Professionals: Therapists, Social Workers, Psychologists, and ADHD Coaches

### 4.6.1 More Than Medication

Successfully managing ADHD throughout life requires more than just medication—it is about integrating pills and skills. Effective treatment involves a combination of structure, support, and strategies tailored to an individual's unique needs. ADHD management requires a multifaceted approach that addresses the diverse challenges individuals face in different aspects of life. The landmark Multimodal Treatment Study of Children with ADHD (MTA Study) provided critical insights into the effectiveness of various treatment strategies. Findings from the study demonstrated that medication alone was not as effective as a multimodal treatment approach in managing ADHD symptoms and improving overall functioning (Molina et al. 2009). Early implementation of this approach has been shown to reduce the risk of developing co-occurring conditions, such as anxiety, depression, and substance use disorders. Furthermore, comprehensive and collaborative care also contributes to better long-term outcomes including improved self-regulation, academic and career success, and even a reduction in the risk of premature mortality. By viewing ADHD treatment as a team effort, involving medical professionals; speech, occupational, and sensory therapists; social workers; psychologists; and ADHD coaches, individuals can develop the skills and strategies needed to thrive at every stage of life (Martínez-Núñez and Quintero 2019).

### 4.6.2 Social Workers

Social workers play a vital role in the multimodal treatment of ADHD. Their ability to recognize ADHD symptoms and connect individuals with necessary resources and support systems is essential for early diagnosis and effective treatment. ADHD symptoms can be easily overlooked or misattributed to other issues, especially in individuals facing complex life circumstances (Tourjman et al. 2022). When left undiagnosed, these symptoms can significantly hinder a person's ability to succeed in daily life. ADHD can go unrecognized in a variety of environments, including schools, community outreach programs, and family services. It is also commonly missed in mental health clinics, therapy sessions, and addiction treatment programs.

Other settings where ADHD may be overlooked include healthcare facilities, employment, and vocational support programs, and legal or correctional systems. Additionally, eating disorder programs, domestic violence intervention services, crisis pregnancy centers, veterans' services, and geriatric initiatives may meet individuals struggling with undiagnosed ADHD. By collaborating with other professionals, social workers can advocate for further evaluation and ensure clients receive proper support. Master's level clinical social workers can evaluate and diagnose ADHD if they have expertise in this area. They are also adept at recognizing co-occurring conditions like anxiety or depression, helping to differentiate between ADHD symptoms and secondary issues (Van Pelt 2010). Furthermore, social workers are instrumental in helping clients manage daily functioning, develop coping strategies, and address environmental factors that may exacerbate symptoms, ensuring a comprehensive and multimodal approach to care.

### 4.6.3 Psychologists

Psychologists play an integral role in the multimodal treatment of ADHD, conducting evaluations and providing therapy for individuals with ADHD across the lifespan. They can diagnose ADHD in children, adolescents, and adults through clinical interviews, behavioral observations, and standardized testing. In addition to diagnosing, psychologists provide education to individuals, families, and professionals to foster a better understanding of ADHD, its challenges, and effective treatment strategies. Therapy and emotional support are essential components of treatment for ADHD, as they address the cognitive, emotional, and behavioral challenges associated with the disorder. Psychologists may use cognitive behavioral therapy, behavioral-based therapies, mindfulness practices, family and couples therapy, and parent training in their efforts to engage people involved in treatment for ADHD (Solanto 2013; Mitchell et al. 2013).

Children with ADHD often face academic challenges due to difficulties with focus, organization, and task completion. School psychologists can collaborate with teachers and school counselors to develop individualized education plans (IEPs) and Section 504 plans that identify the unique educational needs as well as practical support for classroom success.

### 4.6.4 Therapists

Speech, occupational, and sensory therapists play a unique role in ADHD treatment. They address communication difficulties, sensory processing challenges, and executive functioning issues often associated with the disorder. Through targeted interventions, these therapists help individuals with ADHD improve focus, social skills, and overall daily functioning. Their approaches may include sensory integration techniques, communication strategies, and motor skill development.

Speech therapists specifically help with communication challenges like taking turns, understanding social cues, and expressing thoughts clearly. They can also address difficulties with following instructions, understanding complex language, and interpreting nonverbal cues (Machado-Nascimento et al. 2016). Occupational and sensory therapists use activities to address sensory processing challenges like over-sensitivity to touch or sounds, helping to improve focus and self-regulation. They can also help with improving fine motor skills like handwriting and coordination. Sometimes they will include cognitive skill training like planning, organizing, time management, and task initiation to help with daily activities or they may suggest using visual aids like schedules and checklists to improve organization and task completion (Ianni et al. 2020).

### 4.6.5 ADHD Coaches

"ADHD coaching is increasingly recognized as an important component of multimodal treatment for individuals with ADHD" (Ahmann et al. 2018). ADHD coaches are instrumental in helping individuals with ADHD take action. Unlike mental health therapists who focus on emotional and psychological aspects, ADHD coaches work in a practical, goal-oriented manner to address daily challenges related to organization, time management, procrastination, and motivation. Everything is in real time and accountability is the corner stone.

ADHD coaching is a structured, supportive intervention that complements therapy and medication. It is particularly beneficial for individuals struggling with day-to-day functioning, motivation, and executive functioning skills. Studies show that coaching can lead to measurable improvements in ADHD management. "Two studies of ADHD coaching programs for college students found that participants engaged in more positive thoughts and behaviors, such as taking greater responsibility for ones actions, using goal attainment skills, modulating emotions and increasing positive expectations for performance" (Prevatt and Levrini 2015). Support is focused on creating personalized strategies for organization, planning, and prioritization.

Coaching uses accountability to ensure follow through. Working with an ADHD coach involves improving working memory, task initiation, and self-regulation. Strategies for impulse control and emotional regulation are taught. Coaches help improve self-esteem by helping people identify their strengths and interests to boost confidence and motivation. ADHD coaches use positive reinforcement to combat ADHD-related self-doubt (Ahmann et al. 2018).

## 4.7 Communication Strategies Among Care Teams

ADHD is one of the most controversial neurodevelopment conditions, yet also one of the most treatable. Unfortunately, "there is no magic symptom related cut off point beyond which ADHD absolutely exists and short of which it does not"

(Hinshaw 2018). Therefore, current, relevant, and evidence-based knowledge about ADHD and its impact across the life span is crucial for all involved along the continuum of treatment and intervention. When team members come together from different professions, the baseline of understanding needs to be similar. This would mean ensuring there are minimal gaps in knowledge and approaches to treating ADHD.

Such gaps in professional care do not only pertain to PCPs and other medical professionals. There are gaps for mental health professionals as well in the recognition, evaluation, and treatment of ADHD. Closing these gaps requires increased awareness of seeing ADHD as a public health issue in which we are all responsible. ADHD is often misunderstood and under-treated, particularly in adults, due to historical misconceptions and outdated diagnostic criteria (Faraone et al. 2021). Studies indicate that many PCPs and mental health professionals receive limited ADHD-specific training, leading to inconsistencies in treatment approaches (Hinshaw 2018).

Many clinicians fail to recognize ADHD as a lifelong condition, often overlooking symptoms in women, minorities, and adults due to diagnostic biases (Nigg et al. 2020). Psychologists and therapists, in particular, may focus on comorbid conditions (e.g., anxiety, depression) without addressing core ADHD impairments. Expanding training on neurodevelopment perspectives and emphasizing ADHD as a public health concern can help bridge these gaps (Barkley 2015).

When collaborative care teams for ADHD—including primary care physicians, psychiatrists, nurse practitioners, psychologists, therapists, educators, and social workers—hold differing beliefs and treatment approaches, it can create problems in delivering consistent care. These differences may result in fragmented treatment plans, confusion for patients and families, and reduced adherence to treatment recommendations.

Medical professionals often prioritize pharmacological interventions as the primary treatment for ADHD (Faraone et al. 2021). In contrast, psychologists, therapists, social workers, and ADHD coaches may emphasize behavioral therapy, executive function coaching, and environmental modifications before considering medication (Pelham et al. 2016). Some providers may be cautious about diagnosing ADHD due to concerns about overdiagnosis and stigma. Meanwhile, other providers may argue that delaying diagnosis and treatment increases the risk of academic struggles, emotional dysregulation, and poor long-term outcomes.

These differing perspectives can create barriers to coordinated and effective treatment. Disagreements over medication, behavioral interventions, diagnosis, and treatment goals may lead to inconsistent guidance for patients and families, ultimately reducing treatment success. To minimize these challenges, improving interdisciplinary communication, establishing shared treatment protocols, and involving families in collaborative decision-making are essential.

Providers may find collaboration time-consuming and challenging. However, there is a vital need for an interdisciplinary and multisystem model (McGonnell et al. 2009) in the evaluation and treatment of ADHD. Care teams should advocate for proactive, clear, and consistent communication, mutual understanding of professional roles, strategies to overcome practical barriers to collaboration, ongoing training, and adaptation to new technologies (Bradley-Klug et al. 2013; McGonnell et al. 2009; Salminen et al. 2025).

Open and honest communication is essential for effective collaboration. Teams should be open-minded and share information. When working with families, providers should use strengths based language and avoid jargon. There should be regular discussions and planning to create an environment for shared learning and dialogue (Salminen et al. 2025). Care teams should have consistent and coordinated messaging to ensure that families hear the same message from all professionals. Additionally, there should be simplified contact for families with the interdisciplinary team (McGonnell et al. 2009).

Medical providers tend to have limited knowledge of the diverse roles of mental health and educational professionals such as school psychologists (Bradley-Klug et al. 2013). Additionally, school psychologists report that pediatric medical professionals are often hard to reach. It may be helpful to utilize e-health technologies for communication and collaboration. It is also important to coordinate collaborative ongoing professional development.

Professionals collaborating and training together in professional associations such as the American Professional Society of ADHD and Related Disorders (APSARD) can help bridge this training and perspective gap. PCPs should be aware that they may have differing views on child development and ADHD in comparison to other fields such as behavioral healthcare. Building relationships through joint training may be one way to improve consistency in training and communication between professions. PCPs should be aware that behavioral parent training (BPT), lifestyle interventions addressing diet and sleep, and school supports are effective adjuncts to medication when treating ADHD and related disorders.

## 4.8 Engagement of Schools, Families, and Community Support

The importance of collaboration and engagement cannot be overstated as providers tend to work in silos. There is a critical need for effective collaboration (Salminen et al. 2025). ADHD is a disorder of self-regulation that impacts multiple domains of life functioning. The diagnosis of ADHD stems from difficulties with attention, inhibition, and cognitive control that are significant enough to cause functional impairment. This functional impairment tends to impact multiple life domains including school, work, social, family, and community. Evidence-based treatment of ADHD must include consideration and potential support in one or more domains of life functioning.

### 4.8.1 Engagement of Schools

There are many educational professionals that can provide important information about a child's academic, emotional, behavioral, and social functioning. Not only can they provide essential information that will be extremely helpful when you are conducting your diagnostic evaluation, but they can assist with treatment implementation and adherence. These professionals include school psychologists, school social workers, school counselors, school teachers, and school nurses.

There are different service delivery frameworks that can be utilized when supporting children with ADHD in the public schools. Supports can be provided through a Multi-Tiered System of Supports, Universal Design for Learning (UDL), Differentiated Instruction (DI), and Specially Designed Instruction (SDI). The first three are relevant to all students. MTSS, UDL, and DI are components of universal instruction that help identify student needs, target interventions, and monitor progress. The focus is on prevention, early intervention, and flexible presentation to enhance student achievement.

The focus on flexible presentation and accommodations can be provided, if necessary, through a Section 504 accommodation plan (Section 504 of the Rehabilitation Act 1973). However, it is important to note that Section 504 is *not* SDI or special education. SDI is delivered through special education as outlined in the IDEA (Individuals with Disabilities Education Act 2004). A student who qualifies for an individualized education program/plan (IEP) is determined to have an educational *disability* in *need* of special education and related services. *Both* of these requirements must be met to qualify for special education, and this decision is made by the school team after conducting an appropriately comprehensive evaluation in all areas of suspected disability.

*Some* students with ADHD may demonstrate significant enough educational impairment to warrant the need for a special education IEP. *Many* students with ADHD may demonstrate enough educational impairment to warrant the accommodations via a Section 504 plan. *However*, not all students with ADHD will demonstrate a need for special education or 504 accommodations. Some students with ADHD can be sufficiently served through regulation education with the supports of MTSS, UDL, and DI.

School psychologists can serve as an excellent resource for the PCP when conducting a medical evaluation for ADHD. School psychologists have access to students during the school day including access to academic and behavioral data. They may have already conducted a psychoeducational evaluation of the student. If the school psychologist conducted a psychoeducational evaluation, they may have cognitive, academic, social, emotional, behavioral, and/or adaptive information on the student that may improve diagnostic precision and lead to targeted medical and behavioral recommendations.

#### 4.8.1.1 Academic Comorbidities

"ADHD adversely affects long-term academic outcomes" (Arnold et al. 2020). It is well-established that the diagnosis of ADHD carries a very high risk of being diagnosed with a second and third comorbid condition (Barkley 2015). This may be because ADHD is marked by significant challenges in executive function/self-regulation, or what Leonard Koziol (Koziol et al. 2013) terms cognitive control. Koziol suggests that challenges in cognitive control may be related to differences in the development of subcortical structures such as the cerebellum and basal ganglia. As a result, ADHD may very well be a foundational disorder (Brown 2017). It is no wonder that ADHD often co-occurs with other neurodevelopmental disorders such as learning disabilities (Pennington et al. 2019).

#### 4.8.1.2 Reading

Dyslexia is a reading disorder that impacts word level reading skills such as phonics and reading fluency. It is estimated that the bidirectional comorbidity of both disorders is 25–40% (Liu et al. 2024). If you include subthreshold/subclinical cases, it may very well be equal to or greater than 50%. The two best indicators of dyslexia are deficits with phonological processing and rapid automatized naming (RAN).

Many individuals with dyslexia present with attention problems; conversely, many individuals with ADHD present with reading problems related to phonological processing and RAN (Cutro 2023). With this knowledge, it should be standard practice to screen for reading difficulties during an ADHD evaluation through a formal interview or screening protocol. A screening such as the Wide Range Achievement Test—Fifth Edition (WRAT-5) could be very useful so that a comorbid learning disorder is not missed.

#### 4.8.1.3 Written Language

Disorders of written language comprise the greatest academic comorbidity with ADHD. This is because the etiology of a written language disorder may be more multifaceted than reading and math. For example, written language problems can be due to difficulties with motor planning and execution, language processing, phonological processing, orthographic processing, and/or executive function. In many individuals with ADHD, written language difficulties may be caused by a combination of factors, meaning that multiple evaluations and interventions may be warranted. "Dysgraphia is common at all ages in children and adolescents with ADHD" (Mayes et al. 2019).

As an example, there is a documented comorbidity between ADHD and developmental coordination disorder. This can be identified by a quick interview or screening protocol, and then a referral can be sent for an evaluation with an occupational therapist. A free screening tool such as the Developmental Coordination Disorder Questionnaire (DCDQ) could be very useful for the PCP.

#### 4.8.1.4 Mathematics

The link between mathematics disorders and ADHD is established but less understood in comparison to reading. The science of math disorders is not as advanced as the science of reading disorders. Nonetheless, many individuals with ADHD struggle with math. They may struggle with number sense, math fact fluency, math calculation skills, and math problem solving. Depending on the profile, an individual with underlying cognitive processing and executive weaknesses may struggle to learn and perform in math. For academic concerns, screening may be as simple as requesting report cards from the school or family when conducting an ADHD evaluation.

### 4.8.2 Engagement of Families

Family engagement may be the most important part of treatment and medication adherence. ADHD still carries a stigma. Some individuals believe that ADHD means you are lazy. Parents can sometimes feel guilt, believing that they caused ADHD with bad parenting. This is not supported by research. ADHD is a largely genetic neurodevelopmental disorder. It is highly unlikely that a parent caused their child's ADHD; however, behavioral parent training (BPT) is an effective adjunctive intervention for treating ADHD. This intervention can help to improve family functioning and ameliorate the intensity of ADHD.

Vigorous exercise, quality sleep, and nutrition are also factors that can improve health and serve to reduce the intensity of ADHD. Sleep quality is an important area to screen when working with families. There is a well-established comorbidity between ADHD and sleep problems (Barkley 2015). There may be structural reasons such as sleep apnea, difficulties with self-regulation that impact sleep onset and the sleep-wake cycle, racing thoughts related to anxiety, and other related difficulties such as restless limb syndrome. Addressing underlying sleep difficulties can improve self-regulation and attentional functioning, reducing the intensity of ADHD.

### 4.8.3 Community Support

The truth is that many medical professionals, including nurse practitioners, do not feel sufficiently trained to evaluate, diagnose, and/or treat ADHD. This is why collaboration with community providers is often essential. There are medical and behavioral providers that can play a role in helping the nurse practitioner with differential diagnosis and treatment planning. This may include consultation and collaboration with psychologists, social workers, speech language pathologists, occupational therapists, and physical therapists. Remember that for ADHD, comorbidity is the norm and not the exception.

PCPs should be aware of the advocacy and support groups that exist for ADHD. These organizations include Children and Adults with Attention-Deficit/Hyperactivity Disorder (CHADD), ADDitude, and Understood—for learning and thinking differences. You may also want to refer to professional organizations such as the American Psychological Association, the National Association of School Psychologists, and state associations such as the Florida Association of School Psychologists.

It is no wonder that when experts talk about ADHD, it is often followed by the words *and related disorders*. PCPs can no longer treat ADHD in isolation as if it exists in a vacuum as a clean categorical entity. ADHD is real but the human brain is complex. The ADHD phenotype likely represents the tip of the iceberg. With this knowledge in mind, PCPs will no longer treat ADHD in isolation but will use their knowledge to screen for related conditions, refer to specialists, collaborate with providers, and monitor beyond a static point to ensure the best patient care and prognosis.

## References


Ahmann E, Tuttle LJ, Saviet M, Wright SD. A descriptive review of ADHD coaching research: implications for college students. J Postsec Edu Disab. 2018;31(1):17–39.

American Association of Colleges of Nursing. Interprofessional partnerships. n.d. https://www.aacnnursing.org/Portals/42/AcademicNursing/pdf/Interprofessional-Partnerships.pdf

American Psychiatric Association. Diagnostic and statistical manual of mental disorders. 5th ed., text rev., DSM-5-TR. Washington, DC: American Psychiatric Publishing; 2022.

Arnold LE, Hodgkins P, Kahle J, Madhoo M, Kewley G. Long-term outcomes of ADHD: academic achievement and performance. J Atten Disord. 2020;24(1):73–85. https://doi.org/10.1177/1087054714566076.

Barkley RA, editor. Attention-deficit hyperactivity disorder: a handbook for diagnosis and treatment. 4th ed. New York: Guilford Press; 2015.

Bradley-Klug KL, Jeffries-DeLoatche KL, St. John Walsh A, Bateman LP, Nadeau J, Powers DJ, Cunningham J. School psychologists' perceptions of primary care partnerships: implications for building the collaborative bridge. Adv School Ment Health Promot. 2013;6(1):51–67. https://doi.org/10.1080/1754730X.2012.760921.

Brown TE. Outside the box: rethinking ADD/ADHD in children and adults: a practical guide. Washington: American Psychiatric Publishing; 2017.

Casey T, O'Neill H, O'Neill B. Adult attention deficit hyperactivity disorder clinic: an interprofessional collaboration. J Am Pharm Assoc. 2020;60(2):273–7.

Cropley S, Duggleby W, Cooper D. Care coordination and the essential role of the nurse. Creat Nurs. 2013;19(2):76–82.

Cutro NM. The diagnostic intersection of ADHD and dyslexia: differential diagnosis, comorbidity, and implications for MTSS and school psychology; 2023. https://ecommons.luc.edu/luc_soecap/20.

Davis CC, Claudius M, Palinkas LA, Wong JB, Leslie LK. Putting families in the center: family perspectives on decision making and ADHD and implications for care. J Atten Disord. 2012;16(8):675–84.

Faraone SV, Banaschewski T, Coghill D, Zheng Y, Biederman J, Bellgrove MA, Newcorn JH, Gignac M, Al Saud NM, Manor I, Rohde LA, Yang L, Cortese S, Almagor D, Stein MA, Albatti TH, Aljoudi HF, Alqahtani MMJ, Asherson P, Wang Y. The World Federation of ADHD international consensus statement: 208 evidence-based conclusions about the disorder. Neurosci Biobehav Rev. 2021;128:789–818. https://doi.org/10.1016/j.neubiorev.2021.01.022.

Hinshaw SP. Attention deficit hyperactivity disorder (ADHD): controversy, developmental mechanisms, and multiple levels of analysis. Annu Rev Clin Psychol. 2018;14:291–316. https://doi.org/10.1146/annurev-clinpsy-050817-084917.

Ianni L, Mazer B, Thomas A, Snider L. The role of occupational therapy with children with attention-deficit hyperactivity disorder: A Canadian national survey. J Occup Ther Sch Early Interv. 2020;14(6):1–22.

Individuals with Disabilities Education Act, 20 U.S.C. §1400 et seq; 2004. https://sites.ed.gov/idea/statute-chapter-33/subchapter-i/1400
Kokorelias KM, Gignac MAM, Naglie G, Cameron JI. Towards a universal model of family centered care: A scoping review. BMC Health Serv Res. 2019;19:564.
Koziol LF, Budding DE, Chidekel D. ADHD as a model of brain-behavior relationships. New York: Springer; 2013.
Lasky AK, Weisner TS, Jensen PS, Hinshaw SP, Hechtman L, Arnold LE. ADHD in context: young adults' reports of the impact of occupational environment on the manifestation of ADHD. Soc Sci Med. 2016;161:160–8.
Liu W, Hong T, Wang J, Zhang L, Kang L, Wang C, Shu H, Wang Y. What factors contribute to reading in ADHD: examining reading-related skills in children with ADHD only/comorbid developmental dyslexia. J Atten Disord. 2024;28(2):201–10. https://doi.org/10.1177/10870547231210287.
Machado-Nascimento N, Melo e Kümmer A, Lemos SM. Speech-language pathology findings in attention deficit hyperactivity disorder: A systematic literature review. CoDAS. 2016;28(6):833–42. https://doi.org/10.1590/2317-1782/20162015270.
MacLeod K, Lewis A, Robertson C. Rethinking the individualized education plan process: voices from the other side of the table. Disabil Soc. 2017;32(3):381–97.
Martínez-Núñez B, Quintero J. Update the multimodal treatment of ADHD (MTA): twenty years of lessons. Actas Esp Psiquiatr. 2019;47(1):16–22.
Matza LS, Paramore C, Prasad M, Stewart KD, Khan SA. Barriers to medication adherence in ADHD: A systematic review. J Atten Disord. 2022;26(5):601–12.
Mayes SD, Breaux RP, Calhoun SL, Frye SS. High prevalence of dysgraphia in elementary through high school students with ADHD and autism. J Atten Disord. 2019;23(8):787–96. https://doi.org/10.1177/1087054717720721.
McGonnell M, Corkum P, McKinnon M, MacPherson M, Williams T, Davidson C, Jones DB, Stephenson D. Doing it right: an interdisciplinary model for the diagnosis of ADHD. J Can Acad Child Adolesc Psychiatry. 2009;18(4):283–6.
Mitchell JT, McIntyre EM, English JS, Dennis MF, Beckham JC, Kollins SH. A pilot trial of mindfulness meditation training for ADHD in adulthood: impact on core symptoms, executive functioning and emotional dysregulation. J Atten Disord. 2013;20(10):1–16.
Molina BSG, Hinshaw SP, Swanson JM, Arnold LE, Vitiello B, Jensen PS, Epstein JN, Hoza B, Hechtman L, Abikoff HB, Elliott GR, Greenhill LL, Newcorn JH, Wells KC, Wigal T, Gibbons RD, Hur K, Houck PR, MTA Cooperative Group. The MTA at 8 years: prospective follow-up of children treated for combined-type ADHD in a multisite study. J Am Acad Child Adolesc Psychiatry. 2009;48(5):484–500. https://doi.org/10.1097/CHI.0b013e31819c23d0.
National Council of State Boards of Nursing. Nurse Practice Act, rules & regulations. n.d. https://www.ncsbn.org/nurse-practice-act.htm
Nigg J, Sibley M, Thapar A, Karalunas S. Development of ADHD: etiology, heterogeneity, and early life course. Annu Rev Dev Psychol. 2020;2(1):559–83. https://doi.org/10.1146/annurev-devpsych-060320-093413.
Pelham WE, Fabiano GA, Massetti GM. Evidence-based assessment and treatment of attention-deficit/hyperactivity disorder. J Clin Child Adolesc Psychol. 2016;45(3):247–62.
Pennington BF, McGrath LM, Peterson RL. Diagnosing learning disorders: from science to practice. 3rd ed. New York: Guilford Press; 2019.
Prevatt F, Levrini A. ADHD coaching: A guide for mental health professionals. Washington, DC: American Psychological Association; 2015.
Ronis SD, Slawek DE, Slawek PA. Patient-centered medical home and family burden in attention-deficit hyperactivity disorder. J Dev Behav Pediatr. 2015;36(2):127–36.
Salminen J, Laasanen M, Leinonen L, Karukivi M, Vornanen R, Alin M, Kyttälä M. Professional collaboration for children and adolescents with neurodevelopmental disorders: a scoping review. Eur J Spec Needs Educ. 2025;40(1):16–34. https://doi.org/10.1080/08856257.2024.2314422.

Schatz NK, Fabiano GA, Cunningham CE, et al. Systematic review of patients' and parents' preferences for ADHD treatment options and processes of care. Patient. 2015;8:483–97. https://doi.org/10.1007/s40271-015-0112-5.

Section 504 of the Rehabilitation Act of 1973, 29 U.S.C. §794; 1973. https://www.dol.gov/agencies/oasam/centers-offices/civil-rights-center/statutes/section-504-rehabilitation-act-of-1973

Solanto M. Cognitive-behavioral therapy for adult ADHD: targeting executive dysfunction. New York: Guilford Press; 2013.

Tourjman V, Louis-Nascan G, Ahmed G, DuBow A, Côté H, Daly N, Daoud G, Espinet S, Flood J, Gagnier-Marandola E, Gignac M, Graziosi G, Mansuri Z, Sadek J. Psychosocial interventions for attention deficit/hyperactivity disorder: A systematic review and meta-analysis by the CADDRA guidelines work group. Brain Sci. 2022;12(8):1023. https://doi.org/10.3390/brainsci12081023.

Van Pelt J. Adult ADHD—hidden diagnosis. Social Work Today. 2010;10(3):14.

Weissenberger S, Ptacek R, Erman A, Schonova K, Raboch J, Goetz M. ADHD, lifestyles and comorbidities: A call for a holistic perspective—from medical to societal intervening factors. Front Psychol. 2017;8:454. https://doi.org/10.3389/fpsyg.2017.00454.

Wolraich ML. It takes a full-service village to treat children with ADHD. Pediatrics. 2016;138(6) https://doi.org/10.1542/peds.2016-2928.

# 5 Diagnostic Formulation of ADHD

Kelsey Doyle and Kate Pfeiffer

## 5.1 Components of an Initial Diagnostic Evaluation for ADHD

ADHD is a chronic condition with overlapping symptoms for many other psychiatric disorders. Symptoms are also highly impacted by environmental stimuli, perceived individual stressors experienced by the individual. Comprehensive and careful clinical assessment is essential for accurate diagnosis of ADHD. This involves gathering a thorough psychiatric history from the individual and other sources, evaluation of current functioning, and a careful review of the DSM-5-TR criteria to determine whether the individual meets the diagnostic threshold.

Before reviewing the specific DSM-5 TR criteria, there are several considerations that influence an accurate diagnosis, such as developmental milestones, family dynamics, environmental observations, and potential comorbidities. These biopsychosocial elements provide critical context for understanding how ADHD may present across different ages and settings, ensuring a more informed, accurate, and individualized diagnostic formulation.

### 5.1.1 Completing an Assessment Through Bio-psycho-social Framework

The following is a framework to consider when forming comprehensive history: (American Academy of Pediatrics 2019) What presenting client experiences overlap with core symptoms of ADHD? (American Psychiatric Association 1980) What

K. Doyle (✉) · K. Pfeiffer
Quinnipiac University, Hamden, CT, USA
e-mail: kelsey.doyle@quinnipiac.edu; Katherine.Pfeiffer@quinnipiac.edu

© The Author(s), under exclusive license to Springer Nature Switzerland AG 2026

R. J. Kennedy (ed.), *Assessment & Treatment of ADHD*,
https://doi.org/10.1007/978-3-032-24171-9_5

historical and current aspects of the client's biological, psychological, and social factors may contribute to the current symptom picture?

When assessing core symptoms of ADHD, it is important to identify specific symptom characteristics and how they present. Clinicians should explore whether the symptoms occur across different settings or are limited to specific situations; this helps to differentiate between environmental triggers and pervasive concerns. It is also crucial to establish the age of symptom onset, as early emergence of symptoms can indicate other developmental or chronic issues, whereas the DSM ADHD diagnostic criterion requires symptom onset before age 12 (American Psychiatric Association 2022). The nature of the symptoms should also be explored in detail: Are there precipitating factors that worsen the presenting symptoms? Anything that helps to relieve symptoms? Do symptoms significantly disrupt age-appropriate functioning and routine activities such as school, home life, or peer relationships? See Table 5.1 for a list of suggested interview questions for thorough history taking with a biopsychosocial lens, as recommended by children and adults with attention-deficit/hyperactivity disorder (CHADD n.d.).

### 5.1.2 Interviewing Client and Family

Eliciting key signs and patterns of inattention, hyperactivity, and impulsivity can come from direct observation of the client and collateral input from family members, teachers, caregivers, and colleagues (American Academy of Pediatrics

**Table 5.1** Considerations for the Biopsychosocial Assessment in ADHD Diagnosis

| Domain | Biological | Psychological | Social |
|---|---|---|---|
| Developmental | Prenatal and birth history: Were there any complications during pregnancy or delivery? | How does the child manage transitions and developmental tasks (e.g., independence, emotional maturity)? | Are there friendships or social challenges at school or in the community? |
| Medical | Any history of illnesses, injuries, hospitalizations, or chronic conditions? | Are there any signs of co-occurring disorders (e.g., ADHD, anxiety, learning disorders)? | Are medical issues impacting school, social engagement, or family participation? |
| Family | Is there a family history of ADHD, learning disabilities, or psychiatric illness? | How does the child describe their relationship with parents/siblings? | Who lives at home? Any recent changes, separations, or conflicts? |
| Function | Describe current sleep, appetite, energy levels, and physical activity | What emotional or behavioral symptoms interfere with daily functioning?<br>How does the individual cope with stress, disappointment, or conflict? | Participation in activities, sports, clubs, or spiritual groups? |

[AAP] 2019). One of the key DSM diagnostic criteria for ADHD is that symptoms occur in multiple domains/different settings for at least 6 months (APA 2022). Assessing multiple domains explores client functioning at home, school, work, and social settings. Collateral contacts provide essential historical and current information about the age of onset, developmental milestones, and early symptoms that the client may not be able to articulate or recognize. Teachers and school staff can also reflect and report on the client's attention, impulsivity, social relationships, and academic performance. This data may be obtained through direct collateral interviews or a structured screening tool (Table 5.2).

**Table 5.2** Rating scales used to assess ADHD across the lifespan

| Rating scale | Source | Scope | Rater types | Age range | Available languages |
|---|---|---|---|---|---|
| Adult ADHD Self-Report Scale (ASRS 5 Version ) | Freely available | Partially represents ADHD DSM symptoms | Self-report | 18+ | Over 20 different languages |
| SNAP-IV | Freely available from author's website | ADHD DSM symptoms; oppositional defiant disorder; partially represents DSM symptoms for anxiety disorders, mood disorders | Caregiver, teacher | 6–18 | English, Spanish, German, French, Chinese, Italian |
| Vanderbilt Assessment Scale | Commercially available | ADHD DSM symptoms; oppositional defiant disorder; conduct disorder, anxiety, depression, functional impairment links | Caregiver, teacher | 6–12 | English, Spanish |
| Conners Youth Scale Fourth Edition | Commercially available | ADHD DSM symptoms, functional impairment | Caregiver teacher, self-report | | English, Spanish, French |
| Conners Adult ADHD Rating Scales Second Edition (CAARS 2) | Commercially available | ADHD DSM symptoms, functional impairment | Caregiver teacher, self-report | 6–18 | English, Spanish, French |
| ADHD Rating Scale-5 | Commercially available | ADHD symptoms, functional impairment | Caregiver, teacher | 5–17 | English, Spanish |
| Wender Utah Rating Scale (WURS-25) | Freely available | Retroactive assessment of childhood ADHD symptoms in adults | Self-report | 18+ | English, French, Italian, Spanish, Bulgarian, Chinese, Swedish |

### 5.1.3 Supplement with Standardized Tools

While the diagnosis of ADHD is made by clinical interviews, standardized interview tools can provide support and structure throughout the interview process. An example of one such interview tool is the Diagnostic Interview for ADHD, which is a structured interviewing tool to identify symptom frequency, severity, and associated impairment across the lifespan with an interview tool for children and adults (Petterson et al. 2015).

## 5.2 Developmental Considerations

Clinicians should approach ADHD assessment and management with a developmentally informed perspective, recognizing that ADHD is a chronic, lifelong condition. Its symptoms and associated impairments can vary significantly across different stages of development, as well as the factors that influence long-term outcomes and identifying predictors.

### 5.2.1 Early and School Age

While ADHD is diagnosed predominantly during school years, there is a growing trend of identifying symptoms in preschoolers (Lahey et al. 2005). It is important to remember that the behavioral manifestations of ADHD, including short attention span, hyperactivity, and poor inhibitory control, are considered a normal finding in healthy preschool children. An estimated 60% of children who are preschool age diagnosed with ADHD continue to meet criteria of the diagnosis into the school age years, supporting validity of the diagnosis (Lahey et al. 2005). Early diagnosis can be reliable, and not an overidentification of expected behavior for preschoolers.

ADHD is most diagnosed during the school-age years related to the assessment of disruptiveness in the classroom setting or poor academic performance. Children may require more support in the classroom, and caregivers may report difficulties following tasks and directions at home. There may be some irritability related to lack of focus.

### 5.2.2 Adolescence and Adulthood

As adolescents develop, inattentive symptoms may become more prominent, while hyperactivity symptoms may decrease in frequency and intensity or evolve into more "socially acceptable" behaviors. For example, running when being seated is required may become excessive foot-tapping, or frequent bathroom trips. ADHD symptoms may emerge or become more noticeable in adulthood when external structure, such as parental support or rigid school routines, is no longer present during the transition to college or other major life transition or significant routine change. As stressors increase, inattention becomes more impairing and may manifest as struggles with organization, job performance, and difficulty with self-regulation.

### 5.2.3 Family Dynamics

Family dynamics play a critical role in the expression and management of ADHD symptoms in children. Children with ADHD often require more structure, consistency, and emotional regulation support than their neurotypical peers (Claussen et al. 2024). When a family system is under chronic stress, such as financial strain, interpersonal conflict, mental health challenges, or caregiving burden, the necessary supports may be limited or inconsistently applied. As a result, ADHD symptoms can intensify, leading to increased impulsivity, defiance, or emotional outbursts.

Parenting style significantly influences outcomes for children with ADHD. Authoritative parenting is characterized by warmth, clear expectations, and consistent consequences and has been associated with better emotional regulation and reduced symptom severity. In contrast, authoritarian (overly harsh or rigid) or permissive (inconsistent or overly lenient) parenting may worsen attention and behavior problems. Children with ADHD may also misinterpret or struggle to respond to emotional cues, leading to increased family conflict if parents lack strategies tailored to the child's needs (Cherkasova et al. 2013).

## 5.3 DSM 5-TR Diagnostic Criteria

### 5.3.1 Diagnosis and Clinical Presentation

The American Psychiatric Association's (APA) *Diagnostic and Statistical Manual of Mental Disorders,* Fifth Edition, Text-Revision (DSM-5-TR) published in March 2022, describes the ADHD diagnosis as a clear pattern of these main features: (1) inattention and/or hyperactive-impulsive behavior that interferes with daily functioning, (2) symptoms occur in multiple domains/different settings for at least 6 months, (3) some of the impairments and symptoms began in early childhood, and (4) no other psychiatric disorder is better attributed to the symptom presentation. Symptoms do not occur exclusively with a psychotic disorder (e.g., schizophrenia) and cannot be better accounted for by another mental disorder, such as depression, medical conditions, or anxiety disorder, although comorbidity may be present (see Sect. 5.5). Adherence to the specific DSM-5-TR criteria is important to minimize clinician adherence to a heuristic model and miss important exclusion criteria or commonly occurring comorbidity (Choi et al. 2022).

Clients may seek treatment once symptoms cause pronounced interference at school, work, and home. This impairment in multiple areas is an important part of the diagnostic evaluation because it distinguishes ADHD from typically expected attention lapses. The consequences of these symptoms and the history of their initial presentation throughout the lifespan, such as early academic challenges, should be carefully explored. The clinical presentation may be primarily hyperactive-inattentive, primarily inattentive, or a combination of the two and includes a wide range of experiences (APA 2022).

Inattention considers the ability to sustain focus and organization, and these symptoms may become prominent in elementary school around 8–9 years old. This may present as clear difficulty focusing on a task, or complaints of feeling "scattered," disorganized, or forgetful, and result in poor attention and mistakes at home, work, or school. Hyperactivity describes excessive motor activity and impulsivity as hasty or reckless actions without planning. This may look like difficulty sitting still or quiet when expected to do so, feeling restless or even frequent movement (e.g., leg bouncing), excessive talking, and impatience (APA 2022).

## 5.4 ADHD Screening Tools Across the Lifespan

Behavior rating scales for ADHD have been used for many years, starting with the symptoms listed in DSM-III (American Psychiatric Association [APA] 1980). These scales have evolved along with the DSM, containing comparable content to the DSM criteria, and some include comorbid disorder assessments to aid in comprehensive evaluation. Rating scales can be a valuable component of both research and clinical assessment, used in conjunction with a structured diagnostic interview. Observer reports are often included in these tools; collateral information can aid in diagnosis due to diagnostic criteria requiring symptom presence across multiple domains. Rating scales and screening tools should not be used independently to diagnose ADHD but can be particularly useful to establish the presence of core ADHD diagnostic symptoms across settings. Narrow-band rating scales focus exclusively on core ADHD symptoms, and broadband scales include constructs or symptom patterns relevant to ADHD. Several rating scales are available for use across the lifespan and vary in complexity, availability, and validity.

When appropriate, these rating scales are requested to be completed by individuals who regularly interact with the client, such as family members, teachers, or colleagues at work. Standardization of scales helps to monitor progress during treatment and can also help assess comorbidities. These tools can provide structure, evidence-based insight into symptom presence, and severity. Disorders including depression, anxiety, autism spectrum disorder, and learning challenges frequently co-occur with ADHD (Cherkasova et al. 2013). For specific tools, refer to Sect. 5.4.

Scales can also be useful in measuring treatment response through assessment of symptom changes over time to evaluate whether change may be occurring. Several scales have been developed to assess ADHD signs and symptoms across the lifespan, and some include ADHD among a broader spectrum of symptoms (Table 5.1). When selecting a rating scale, its purpose, validity, and intended population should be considered.

### 5.4.1 Rating Scales for Youth

#### 5.4.1.1 Swanson, Nolan, and Pelham-IV (SNAP-IV)

Several commonly used scales for children and adolescents are available. The Swanson, Nolan, and Pelham-IV (SNAP-IV) focuses on ADHD, oppositional defiant disorder (ODD), and other psychiatric disorders in children on 90-question self-report scales (Swanson 1992). For a quicker standardized way to understand the client's experience, the validated SNAP-IV revisions include the SNAP-IV-26 item scale which measures ADHD, ODD symptoms, and the SNAP-IV-18 which measures only ADHD symptoms. The SNAP scales are available online.

#### 5.4.1.2 Vanderbilt ADHD Diagnostic Parent/Teacher Rating Scales

The Vanderbilt ADHD Diagnostic Parent Rating Scale (VADPRS) (Wolraich et al. 2003) and Teacher Rating Scale (VADRTS) (Wolraich et al. 1998) focus on DSM symptoms of ADHD and ODD, conduct disorder (CD), anxiety and depression in children ages 6–12. The caregiver form has 47 items, and the teacher form has 35 items. The National Institute for Children's Health Quality and the American Academy of Pediatrics recommend it, and it is widely used. These assessment forms can be ordered through the American Academy of Pediatrics bookstore and are also freely available online.

#### 5.4.1.3 Conners Rating Scales

The Conners Fourth Edition is the latest version of the widely used Conners Rating Scales (Conners 2022). With a form for caregivers, teachers, and youth, this tool can be obtained for children and adolescents age 6–18. Children ages 8 and older can complete the self-report. This ADHD assessment also screens for oppositional defiant disorder (ODD), conduct disorder (CD), social problems, and executive functioning. The full-length tool ranges from 109–118 items, and the short forms contain approximately 50 items.

#### 5.4.1.4 ADHD Rating Scale-5

The ADHD Rating Scale-5 is an 18-item scale that evaluates all DSM-5 inattentive and hyperactive/impulsive criteria and related functional impairment in children aged 5–17 and contains forms for parent/caretakers and teachers (DuPaul et al. 2016). This is commercially available for purchase.

### 5.4.2 Rating Scales for Adults

Along with the increase in recognition of ADHD in adults, tools used to assess the disorder and monitor treatment outcomes in this population have also increased.

#### 5.4.2.1 Adult ADHD Self-report Scale

Developed with the World Health Organization and the Workgroup on Adult ADHD, the Adult ADHD Self-Report Scale (ASRS) is an 18-item self-report screening tool for adults 18 or older with suspected ADHD based on DSM-5 and ICD-10 (Kessler et al. 2005). An updated version, the ASRS-5, is a six-item self-report screener commonly used as a quick screening aid in both primary care and psychiatric settings, updated to DSM-5 criteria (Ustun et al. 2017).

#### 5.4.2.2 Conners Adult ADHD Rating Scales

The Conners Adult ADHD Rating Scales (CAARS) were originally published in 1998 as a complement to the child and adolescent scales, and the most recent version is the CAARS 2, published in 2023. Intended for individuals 18 years old and older, there are self- and observer reports available in six different formats that vary by length and scope. The full-length screener contains 97 items that include associated screening items for suicidality, self-harm, anxiety, and depression symptoms (Conners et al. 2023). Two shorter forms are also available. These scales are commercially available.

#### 5.4.2.3 Wender Utah Rating Scale (WURS)

The Wender Utah Rating Scale (WURS) is a self-report for adults 18 and older designed to retrospectively assess ADHD symptoms and behaviors from childhood (Ward et al. 1993). The original tool contains 60 items, and a 25-question subset of those items is used much more commonly, retitled the WURS-25, was designed solely for ADHD assessment.

### 5.4.3 "Broad Band" Rating Scales

Broadband rating scales are assessment tools that measure behavior across multiple domains, covering a wide range of symptoms to provide a comprehensive overview of the client's experience. They may lack specificity and can be more time-consuming, but the information provided from a broadband scale can help identify other conditions that may co-occur with ADHD and influence individualized and comprehensive treatment.

#### 5.4.3.1 Conners Early Childhood

The Conners Early Childhood (Conners EC) is a broadband assessment tool, completed by caregivers, teachers, and childcare providers to describe children ages 2–6 years old (Conners 2009). Not diagnostically specific, the Conners EC assesses ADHD constructs such as hyperactivity and inattention as well as behavioral, cognitive, emotional, and developmental concerns of early childhood. The full-length tool has 190 items, and abbreviated versions range from 48 to 116 items. The Conners tools are commercially available for purchase.

#### 5.4.3.2 Child Behavior Checklist

The Child Behavior Checklist (CBCL) is a component of The Achenbach System of Empirically Based Assessment (ASEBA) (Achenbach and Rescorla 2001). The ASEBA is a variety of tools used together for assessment of several emotional and behavioral problems. The CBCL is a widely used 113-item caregiver report measure assessing anxiety, depression, self-harm, suicidality, conduct disorder, ADHD, and other issues in children and adolescents.

#### 5.4.3.3 Brown Executive Function/Attention Rating Scales

The Brown Executive Function/Attention Rating Scales (BEFARS) is a research and clinical tool to assess executive functioning for ages 3 through adulthood (Brown 2019). Revised from the previous Brown Attention-Deficit Disorder Scales (BADDS), this instrument has forms for preschool starting at age 3, school-age children, adolescents, and adults, with emphasis on specific areas of executive functioning: emotion, action, memory, focus, effort, and activation (Brown 2019).

## 5.5 Assessment of Comorbidity

Comorbidities may be common, so careful differential diagnosis is important. Mood disorders, anxiety, post-traumatic stress disorder, personality disorders, and obsessive-compulsive disorder can have significant symptom overlap, and the distinction between differential diagnosis and comorbidity is not well defined.

Trauma can also significantly impact a child's attention, behavior, and emotional regulation. These are symptoms that often resemble those of ADHD (Miller 2024). Children who have experienced trauma may display hypervigilance, distractibility, impulsivity, and restlessness, which are common in ADHD but may be responses to chronic stress or a perceived need for safety. Similarly, anxiety can manifest as inattention, fidgeting, and difficulty focusing, especially in structured settings like classrooms.

Sensory processing concerns may also lead to behaviors that mimic hyperactive or inattentive, as children struggle to regulate their responses to environmental stimuli. Additionally, language delays can cause a child to disengage or become frustrated during tasks that require verbal comprehension, leading to misinterpretation of their behavior as defiance or inattention (Table 5.3).

Psychiatric comorbidity is highly prevalent among children and adolescents with ADHD. In a systematic review and meta-analysis of 121 studies involving 39,894 youth with ADHD, the most common psychiatric comorbidities were oppositional defiant disorder (34.7%), behavior disorders (30.7%), anxiety disorders (18.4%), specific phobias (11.0%), enuresis (10.8%), and conduct disorder (10.7%) (Njardvik et al., 2025). Because this review focused specifically on psychiatric disorders, it did not include neurodevelopmental disorders such as Specific Learning Disorder (SLD); however, other studies have estimated that approximately 25%–50% of children with ADHD have a comorbid learning disorder, including impairments in reading, written expression, and mathematics, further contributing to academic and functional impairment (Español-Martín et al., 2023). Obsessive-compulsive disorders (OCD) and tic disorders are associated with ADHD and may be more commonly seen in the pediatric population compared with adults (APA 2022). Of note, there are no exclusion criteria for autism spectrum disorder, so these may also

**Table 5.3** Broadband scales used to assess ADHD and multiple conditions across the lifespan

| | | | | | |
|---|---|---|---|---|---|
| Child behavior checklist The Achenbach System of Empirically Based Assessment (ASEBA) | Commercially available | Hyperactivity, inattention, behavioral, cognitive, emotional, and developmental concerns | Caregiver self-report | 1.5– | English, Spanish, French, Tagalog, Vietnamese, Chinese, American Sign Language, Farsi, Polish, Russian, Urdu |
| Conners Early Childhood (Conners EC) | Commercially available | Hyperactivity, inattention, behavioral, cognitive, emotional, and developmental concerns | Caregiver, teacher | 2–6 | English, Spanish |
| Brown Executive Function/ Attention Rating Scales (BEFARS) | Commercially available | Hyperactivity, inattention, executive functioning | Caregiver, teacher, self-report | 3-adulthood | English, French, & Spanish |

co-occur. ADHD is also associated with increased risk of depression, substance use, anxiety disorders, and sleep disorders, across the lifespan (Faraone et al. 2021). Completing standardized screening tools for ODD, anxiety, and MDD can help cluster the symptom picture to organize differential diagnosis. ADHD is also a risk factor for increased suicide attempts and deaths by suicide (APA 2022).

The presence of diagnostic symptom overlap requires careful attention to the diagnostic hierarchy. A careful, stepwise process to approaching comorbidity ensures that more primary disorders (i.e., medical, mood, or psychotic disorders) are stabilized before diagnosing ADHD. This approach helps avoid misdiagnosis and ensures an accurately targeted individualized treatment approach for co-occurring disorders. Screening tools for high-probability comorbidity can clarify symptom clusters across diagnoses and should be used alongside ADHD screening tools. Such commonly used rating scales include the Child Behavior Checklist (CBCL), Autism Spectrum Quotient (AQ) and RITVO for Asperger's, the Epworth Sleepiness Scale for insomnia Beck Depression and Anxiety Rating Scales, GAD-7, PHQ-9, and Beck scales for anxiety and depression, Mood Disorders Questionnaire (MDQ) for Mood Disorder Questionaire and Rapid Mood Screen for mood and bipolar spectrum disorders, and CAGE-AID AUDIT, CAGE-AID, and DAST-10 for alcohol and drug screening. screen for risky alcohol and substance use.

Functional consequences of ADHD-related impairment can range and may include problems in academic and occupational achievement and stability, difficulty relating to peers, and interpersonal challenges (APA 2022). ADHD may also co-occur with other neurodevelopmental disorders such as learning delays.

## References

Achenbach TM, Rescorla LA. Manual for the ASEBA School-age Forms & Profiles. Burlington: University of Vermont, Research Center for Children, Youth, & Families; 2001.

American Academy of Pediatrics. Clinical practice guideline for the diagnosis, evaluation, and treatment of attention-deficit/hyperactivity disorder in children and adolescents. Pediatrics. 2019;144(4):e20192528. https://doi.org/10.1542/peds.2019-2528.

American Psychiatric Association. Diagnostic and statistical manual of mental disorders. 3rd ed. New York: American Psychiatric Association Publishing; 1980.

American Psychiatric Association. Diagnostic and statistical manual of mental disorders. 5th ed. New York: American Psychiatric Association Publishing; 2022.

Brown TE. Brown Executive Function/Attention Scales. San Antonio (TX):Pearson Assessment; 2019.

CHADD. Children and adults with attention-deficit/hyperactivity disorder. n.d. https://chadd.org

Cherkasova M, Sulla EM, Dalena KL, Pondé MP, Hechtman L. Developmental course of attention deficit hyperactivity disorder and its predictors. J. Can. Acad. Child Adolesc. Psychiatry. 2013;22(1):47–54.

Choi WS, Woo YS, Wang S-M, Lim HK, Bahk WM. The prevalence of psychiatric comorbidities in adult ADHD compared with non-ADHD populations: a systematic literature review. PLoS One. 2022;17(11):e0277175. https://doi.org/10.1371/journal.pone.0277175.

Claussen AH, Holbrook JR, Hutchins HJ, Robinson LR, Bloomfield J, Meng L, Bitsko RH, O'Masta B, Cerles A, Maher B, Rush M, Kaminski JW. All in the family? A systematic review and meta-analysis of parenting and family environment as risk factors for attention-deficit/hyperactivity disorder (ADHD) in children. Prev Sci Off J Soc Prev Res. 2024;25(Suppl 2):249–71. https://doi.org/10.1007/s11121-022-01358-4.

Conners CK. Conners early childhood™. Toronto: Multi-Health Systems; 2009.

Conners CK. Conners manual. 4th ed. Toronto: Multi-Health Systems; 2022.

Conners C, Erhardt D, Sparrow E. CAARS 2: Conners adult ADHD rating scales. Toronto: Multi-Health Systems; 2023.

DuPaul GJ, Power TJ, Anastopoulos AD, Reid R. ADHD rating scale-5 for children and adolescents. In: Checklists, norms, and clinical interpretation. Revised ed. New York: Guildford Press; 2016.

Elia J, Ambrosini P, Berrettini W. ADHD characteristics: I. Concurrent co-morbidity patterns in children & adolescents. Child Adolesc Psychiatry Ment Health. 2008;2(1):15. https://doi-org.libraryproxy.quinnipiac.edu/10.1186/1753-2000-2-15

Español-Martín G, Pagerols M, Prat R, Rivas C, Corrales M, Bosch R, Casas M. The impact of attention-deficit/hyperactivity disorder and specific learning disorders on academic performance in Spanish children from a low-middle- and a high-income population. Frontiers in Psychiatry. 2023;14:1136994

Faraone SV, Banaschewski T, Coghill D, Zheng Y, Biederman J, Bellgrove MA, Newcorn JH, Gignac M, Al Saud NM, Manor I, Rohde LA, Yang L, Cortese S, Almagor D, Stein MA, Albatti TH, Aljoudi HF, Alqahtani MMJ, Asherson P, Atwoli L, et al. The World Federation of ADHD International Consensus statement: 208 evidence-based conclusions about the disorder. Neurosci Biobehav Rev. 2021;128:789–818. https://doi.org/10.1016/j.neubiorev.2021.01.022.

Kessler RC, Adler L, Ames M, Demler O, Faraone S, Hiripi E, Howes MJ, Jin R, Secnik K, Spencer T, Ustun TB, Walters EE. The World Health Organization adult ADHD self-report scale (ASRS): a short screening scale for use in the general population. Psychol Med. 2005;35(2):245–56. https://doi.org/10.1017/s0033291704002892.

Lahey BB, Pelham WE, Loney J, Lee SS, Willcutt E. Instability of the DSM-IV subtypes of ADHD from preschool through elementary school. Arch Gen Psychiatry. 2005;62(8):896–902. https://doi.org/10.1001/archpsyc.62.8.896.

Miller C. Is it ADHD or trauma? Why the symptoms are often confused, and how to avoid a misdiagnosis. New York: Child Mind Institute; 2024.

Njardvik U, Wergeland GJ, Riise EN, Hannesdottir DK, Öst LG. Psychiatric comorbidity in children and adolescents with ADHD: A systematic review and meta-analysis. Clinical Psychology Review. 2025;118:102571.

Petterson R, Söderström S, Nilsson K. Diagnosing ADHD in adults: an examination of the discriminative validity of neruopsychological tests and diagnostic assessment instruments. J Atten Disord. 2015;22(11):1019–31. https://doi.org/10.1177/1087054715618788.

Swanson JM. School-based assessments and interventions for ADD students. Irvine, CA: KC Publishing; 1992.

Ustun B, Adler LA, Rudin C, Faraone SV, Spencer TJ, Berglund P, Gruber MJ, Kessler RC. The World Health Organization adult attention-deficit/hyperactivity disorder self-report screening scale for dsm-5. JAMA Psychiatry. 2017;74(5):520–7. https://doi.org/10.1001/jamapsychiatry.2017.0298.

Ward MF, Wender PH, Reimherr FW. The Wender Utah rating scale: an aid in the retrospective diagnosis of childhood attention deficit hyperactivity disorder. Am J Psychiatry. 1993;150:885–90.

Wolraich ML, Feurer ID, Hannah JN, Baumgaertel A, Pinnock TY. Obtaining systematic teacher reports of disruptive behavior disorders utilizing DSM-IV. J Abnorm Child Psychol. 1998;26(2):141–52. https://doi.org/10.1023/a:1022673906401.

Wolraich ML, Lambert W, Doffing MA, Bickman L, Simmons T, Worley K. Psychometric properties of the Vanderbilt ADHD diagnostic parent rating scale in a referred population. J Pediatr Psychol. 2003;28(8):559–67. https://doi.org/10.1093/jpepsy/jsg046.

# Differential Diagnoses of Comorbid Disorders

# 6

Tiffany D. Gishizky

## 6.1 Introduction

The importance of differential diagnosis and comorbid disorders in ADHD assessment is particularly relevant due to the implications of treatment selection and patient outcomes. Misdiagnosis or failure to properly identify comorbid presentations can lead to suboptimal treatment response, poor clinical outcomes, and possible precipitation or exacerbation of the comorbidity. Conversely, the accurate identification and differentiation of ADHD and its comorbid disorders provides a roadmap for more targeted therapies and effective interventions. This underscores the importance of conducting a comprehensive assessment that differentiates ADHD from other conditions while, at the same time, identifying relevant comorbidities.

As discussed in previous chapters, recent advances in our understanding of ADHD have highlighted the need for nuanced approaches to assessment that consider developmental, cultural, and gender-specific factors. These considerations become particularly important when evaluating comorbid conditions, as the presentation and impact of both ADHD and its comorbidities can vary significantly across different demographic groups and developmental stages (Kooij et al. 2012).

This chapter aims to provide clinicians with a comprehensive framework for approaching the differential diagnosis of ADHD and the assessment of comorbid conditions. We will explore systematic approaches to evaluation, examine specific diagnostic challenges, and discuss evidence-based strategies for distinguishing ADHD from other conditions while identifying relevant comorbidities. Through this discussion, we seek to enhance clinicians' ability to make accurate diagnostic

T. D. Gishizky (✉)
University of San Francisco, San Francisco, California, USA

NPower ADHD, Bordeaux, France
e-mail: tiffany@npoweradhd.com

© The Author(s), under exclusive license to Springer Nature Switzerland AG 2026

R. J. Kennedy (ed.), *Assessment & Treatment of ADHD*,
https://doi.org/10.1007/978-3-032-24171-9_6

determinations and develop appropriate treatment plans that address both ADHD and its commonly associated conditions.

## 6.2 Differential Diagnoses

### 6.2.1 Medical Conditions

Several medical diagnoses can present as or mimic symptoms of ADHD such as inattentiveness, restlessness, or hyperactivity. Examples of these include but are not limited to thyroid disorders, absence seizures, diabetes, sleep disorders, lead poisoning, hearing loss, and anemia (Pearl et al. 2015). It is therefore imperative that a comprehensive medical evaluation is conducted prior to or alongside an ADHD evaluation and should include a full medical history, physical assessment, and laboratory studies (Sadak 2023). This can be accomplished through collaboration with the patient and their primary medical provider to obtain medical clearance for the ADHD assessment and minimize possible misdiagnosis.

### 6.2.2 Neurodevelopmental Disorders

There are several neurodevelopmental disorders that can present with symptoms similar to those of ADHD, making differential diagnosis particularly challenging (Pearl et al. 2015). Autism spectrum disorder (ASD) frequently includes symptoms related to attention difficulties, impulsivity, and apparent hyperactivity, though these behaviors are often a result of different underlying mechanisms compared with ADHD, such as sensory sensitivities or rigidity in routines (Lai et al. 2019). Specific learning disorders can result in apparent inattention, particularly in academic settings, as students struggle to process information or maintain focus on tasks that are especially challenging for them (Sahoo et al. 2015). Language disorders may create the appearance of inattention or impulsivity when children have difficulty following verbal instructions or participating in verbal exchanges. Developmental coordination disorder (DCD) can present with restlessness and apparent hyperactivity due to motor planning difficulties, while also impacting attention during tasks requiring fine motor skills (Lachambre et al. 2021). The complexity of differentiating these conditions is further heightened by their frequent co-occurrence with ADHD, requiring clinicians to determine whether symptoms represent a distinct neurodevelopmental disorder, ADHD, or both conditions simultaneously.

### 6.2.3 Psychological Conditions

Psychological disorders frequently present with symptoms that can be mistaken for ADHD, requiring careful differential diagnosis. Anxiety disorders may manifest as inattention and restlessness when worry interferes with concentration and task initiation/completion. In mood disorders, specifically bipolar disorder, episodes of mania or hypomania can present with features similar to the hyperactive traits of ADHD such as increased activity, impulsivity, as well as racing thoughts and sleep disruption, while depressive episodes may cause difficulties with concentration and task initiation that resemble inattentive presentations (Sanches et al. 2015). Trauma-related disorders are particularly challenging to differentiate, as hyperarousal and difficulty concentrating due to traumatic experiences can appear similar to ADHD symptoms; moreover, children with trauma histories may display hyperarousal that can be mistaken for hyperactivity and impulsivity as manifestations of emotional dysregulation (Szynabski et al. 2011). Obsessive-compulsive disorder can create apparent inattention when intrusive thoughts interfere with task focus, while compulsive behaviors may be mistaken for hyperactivity. Conversely, what may appear to be checking behaviors associated with OCD may in fact be repetitive task engagement secondary to poor recall present in ADHD (Abramovitch et al. 2013) The diagnostic complexity is further compounded by the fact that these psychological conditions not only mimic ADHD but also frequently co-occur with it, requiring clinicians to carefully evaluate the temporal relationship and qualitative nature of symptoms within each disorder's unique context.

### 6.2.4 Environmental Factors

Environmental factors can create behavioral presentations that closely resemble ADHD symptoms. This informs the clinician of the need for careful contextual assessment in order to determine an accurate diagnosis. Chronic stress within the family system, such as conflict, a chaotic environment, or financial instability, may lead to difficulty concentrating, emotional dysregulation, and behavioral challenges that mirror ADHD traits (Jaffee et al. 2012). Academic incongruities, where instruction is either excessively or insufficiently challenging, can result in apparent inattention, disengagement, and/or classroom disruption that may be mistaken for ADHD. Sleep deprivation is a growing issue among not only children and adolescents but also in adults due to ease of access to and excessive use of electronic devices and demanding schedules. This can manifest as inattention and difficulty with sustained focus, irritability, and hyperactivity. Nutritional factors, including poor diet quality, irregular meal timing, or excessive caffeine consumption, may impact attention and behavior in ways that resemble ADHD. Inadequate physical activity or prolonged periods of sedentary behavior may result in restlessness and difficulty maintaining attention, especially in young children who require regular movement opportunities for optimal functioning.

## 6.3 Common Comorbid Disorders

### 6.3.1 Externalizing Disorders

Externalizing disorders frequently co-occur with ADHD, creating complex clinical presentations that require comprehensive assessment and targeted intervention approaches. These conditions initiate in childhood but can persist into adulthood and are often reported by either the patient or support persons (parents, teachers, spouse) to be the most problematic and interfering issues. This is usually due to the fact that they are the most observable or that cause the most conflict. Oppositional defiant disorder (ODD), characterized by persistent patterns of argumentative behavior, defiance, and emotional dysregulation and conduct disorder (CD), a more severe disruptive behavior disorder, is marked by persistent patterns of rule-breaking, aggression, and violation of others' rights, both of which often lead to more significant impairment in social and academic functioning than ADHD alone (National Academies of Sciences, Engineering, and Medicine et al. 2015). It should be of note that the diagnosis of both ODD and CD in the absence of an ADHD diagnosis occurs more frequently in children belonging to ethnic and racial minority groups (Fadus et al. 2019). It is therefore important to be aware of any potential unconscious bias that may impact a diagnosis and increase diligence in evaluating for the presence of a neurodivergent condition such as ADHD. Disruptive mood dysregulation disorder (DMDD), a more recently recognized condition, and intermittent explosive disorder are both highly co-occurring in children and adults with ADHD and is characterized by severe, chronic irritability and frequent temper/anger outbursts that exceed the emotional dysregulation typically seen in ADHD (National Academies of Sciences, Engineering, and Medicine et al. 2015; Gelegen and Tamam 2018). These comorbid conditions not only complicate the diagnostic process but also significantly impact prognosis and treatment planning, as the presence of externalizing comorbidities often predicts poorer outcomes and necessitates more intensive, multimodal intervention approaches.

### 6.3.2 Internalizing Disorders

The primary internalizing disorders that can present with symptoms that closely resemble ADHD are anxiety and depression. Significant symptom overlap occurs between ADHD and episodic mood disorders and can manifest as difficulty concentrating, restlessness, and problems with task completion—all hallmarks typically associated with ADHD. For instance, someone experiencing anxiety might appear distractible because they're preoccupied with worried thoughts, while depression can cause mental fog and executive function difficulties that mirror ADHD's effects on organization and motivation. The key difference often lies in the underlying mechanisms: while ADHD symptoms typically stem from differences in executive function and attention regulation, internalizing disorders' similar presentations usually result from emotional distress or mood disruption impacting cognitive performance. Another factor to closely consider during evaluation is the episodic nature of symptom presentation related to depression and anxiety. Despite some presentations being protracted in nature, there is often a period of remission in which all of the associated and similar symptoms to ADHD remit as well. For an individual with ADHD, to some degree of severity, these traits are persistent in the absence of mood episodes. This overlap, however, highlights the importance of comprehensive psychological evaluation to ensure accurate diagnosis and appropriate treatment.

### 6.3.3 Developmental Disorders

Autism spectrum disorder (ASD) shares several traits and in fact is highly comorbid with ADHD with research pointing to the difficulty in clinically distinguishing the two (Sokolova et al. 2017). Key similarities include difficulty with social cues, hyper-fixations, sensory sensitivity, and overstimulation, and several behaviors of ASD can appear as hyperactivity (Dickerson Mayes et al. 2012). .

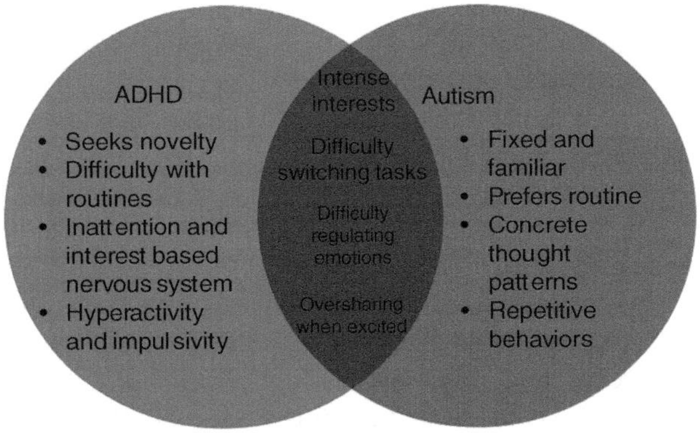


An example of how to distinguish an overlapping trait such as difficulty with social cues is understanding how each of the diagnoses contribute uniquely to this impairment. For example, for ADHD, the difficulty with sustained focus leads to missing parts of a conversation and therefore missing context compared to a difficulty in identifying nuance and intonation as part of the unique neuro-processing associated with ASD.

Language disorders can lead to apparent inattention in academic or social settings as children struggle to process and respond to verbal information, potentially being mistaken for ADHD-related distraction. Similarly, learning disabilities like dyslexia or dyscalculia may cause children to appear inattentive or avoidant during specific tasks, when in reality they're struggling with the fundamental processing demands of those activities (Gnanavel et al. 2019). Intellectual disabilities can also present with attention difficulties and behavioral regulation challenges that superficially resemble ADHD symptoms, though these arise from broader cognitive differences rather than specific attention-regulation issues.

## 6.4 Assessment Strategies for Comorbid Conditions

### 6.4.1 Comprehensive Evaluation Protocols

Best practice for evaluating comorbid conditions that may mimic ADHD involves implementing a comprehensive, multi-modal assessment approach. Clinicians should conduct thorough clinical interviews that explicitly explore symptoms of anxiety, depression, trauma, learning disabilities, autism spectrum disorder, and other neurodevelopmental conditions, being attentive to subtle differences in symptom presentation, triggers, and contexts. Often, due to time limitations, the full evaluation for ADHD will require additional time beyond the general psychiatric evaluation and having the clear expectation that a comprehensive evaluation for ADHD will usually take 2 to 3 hours over the same amount of appointments is important.

Sleep assessments should be included, as sleep disorders can significantly impact attention and behavior; however, insomnia is also a common comorbidity of ADHD itself due to the difficulty or inability to quiet mental and/or physical restlessness.

Importantly, clinicians should adopt a developmental perspective, considering how symptoms have evolved over time and whether they align with typical ADHD trajectories or suggest alternative explanations. This thorough approach helps avoid misdiagnosis and ensures treatment plans address the true underlying conditions.

### 6.4.2 Cognitive Testing

When cognitive testing is utilized, assessment should include measures of sustained attention, executive functioning, processing speed, and working memory. Instruments may include continuous performance tests (CPTs), Stroop tasks, processing speed measures, and working memory assessments such as Digit Span and story memory tasks from the CMS or WMS, which may detect auditory working memory deficits commonly observed in individuals with ADHD, including those with average to very superior intellectual ability (Kennedy et al., 2019; Brown, 2013). Additional tools can include broad-spectrum measures that screen across multiple domains of functioning, not just ADHD-specific symptoms (Liu et al. 2021; Raiker et al. 2018). These tools incorporate multi-modal reporting across settings including self-reporting as well as teachers and parents in child and adolescent evaluation and spouse or significant others for adults.

Cognitive testing help evaluate specific cognitive domains affected by ADHD but also help differentiate between executive function deficits characteristic of ADHD versus those arising from other conditions. It is important to note, however, that while several cognitive tests are high in specificity and validity of the domains being measured, they do not provide a quantitative confirmation of an ADHD diagnosis, and any results should therefore be evaluated with this in mind (Marshall et al. 2018).

## 6.5 Case Examples and Clinical Applications

### 6.5.1 Complex Case I

James, a 27-year-old graduate student, sought evaluation after struggling with academic performance despite his high intelligence. Since childhood, he had difficulty with sustained attention, organization, and task completion. He was easily distracted, often lost important items, and struggled to meet deadlines. These symptoms had been present consistently throughout his life, causing significant impairment in school and later in college.

What complicated his clinical picture was a pattern of mood episodes that began in his late teens. Periodically, James would experience 7–10 day stretches of decreased sleep need (sleeping only 3–4 h while feeling energized), racing thoughts, increased goal-directed activity, and excessive spending. During these episodes, his already-present distractibility would significantly worsen but with a different quality—described as his "mind working too fast" rather than being pulled by external stimuli. Following these episodes, James would often crash into periods of low energy, increased sleep, and profound guilt about tasks left incomplete during his "high" periods.

James had initially been diagnosed with ADHD in college and prescribed stimulant medication, which helped his focus but occasionally seemed to trigger mood elevation. When a comprehensive evaluation was conducted, including mood charting, family history assessment (revealing bipolar disorder in his father), and careful differentiation between chronic attention difficulties and episodic mood-related cognitive changes, clinicians identified both ADHD and Bipolar Disorder Type II.

His treatment plan required careful coordination, starting with mood stabilization through a combination of lamotrigine and lifestyle modifications focused on sleep hygiene. Once his mood was relatively stable for 3 months, a careful trial of non-stimulant ADHD medication (atomoxetine) was introduced, along with cognitive-behavioral therapy specifically targeting executive function skills and mood monitoring. Regular check-ins with both his psychiatrist and therapist allowed for ongoing assessment of both conditions and medication adjustments as needed.

### 6.5.2 Complex Case II

Alex is a 30-year-old trans male (assigned female at birth) that has been engaged in psychiatric treatment for the past 3 years for the diagnoses of ADHD, PTSD, and depression after a full clinical evaluation of his history by the same provider. Alex has been managing his ADHD well and consistently for the past several years and has not required medications for PTSD or depression symptoms. Besides taking stimulant medication for ADHD symptoms, he is receiving ongoing hormone therapy with testosterone and has been generally stable for the past 3 years.

At a recent 3-month psychiatry follow-up appointment, Alex states that both his primary care provider and therapist diagnosed him 2 months ago with bipolar disorder due to experiencing what appeared to be a manic episode. Alex described a period of about 3 weeks of uncharacteristic behaviors, such as hyper sexuality with multiple partners and feeling generally more impulsive, all symptoms that would indicate he did in fact experience a manic episode due to the duration and acuity of symptoms. The challenge facing Alex's psychiatry provider was…why now? There had never been any previous history of manic or hypomanic episodes at initial or ongoing assessments, particularly in light of treatment with stimulant medications.

The provider inquired with Alex what else had been going on in his life around the time of symptom onset: work or relationship stress, changes in sleep, or anything that could have contributed to these behaviors, but Alex gave no indicators that anything had changed. Then a very important question was asked, have there been any changes to the testosterone treatment in the past few months? Alex reported that, in fact, his primary care provider had increased his testosterone dose about a week before symptom onset. While this could not be identified as causal, it was certainly correlative based on what is known about the relationship between testosterone and the hyperactive/impulsive traits of ADHD.

This correlation was discussed and that what likely occurred was a surge in these symptoms that mimicked a manic episode. Rather than initiate a new medication indicated for a bipolar diagnosis, the stimulant dose was adjusted to accommodate for the increased testosterone and Alex's symptoms stabilized with demonstrated efficacy at 6- and 9-month follow-up encounters.

### 6.5.3 Incorporation of Findings and Strategies for Integrated Management

Screening for ADHD across the lifespan should be as prioritized at initial evaluation as presentation of mood disorders, psychotic disorders, and environmental or psychosocial factors are. This is due to the evidence of how significant the symptom burden and impairments secondary to ADHD place on an individual throughout life and in all settings (Lee et al. 2023). Therefore, determining a comprehensive diagnostic picture cannot overlook or treat as an afterthought the inclusion of an initial ADHD screening and subsequent thorough evaluation if suspected.

Having a comorbid diagnosis along with ADHD does not mean that this should be prioritized when determining treatment planning. As discussed earlier in this chapter, there are several symptoms that are present as a result of an ADHD diagnosis itself and would therefore benefit from placing the highest emphasis on ADHD management first. For example, a significant amount of individuals with ADHD struggle with secondary or situational anxiety, sleep and circadian rhythm disruptions, and negative sense of self, all as a result of the persistence of impairment and a lack of lived experience that gives the individual a perspective of what it is like to not live with ADHD. Even in the presence of a determined diagnosis of depression or anxiety, by nature of the rapidity of response to treatment of ADHD (4–8 weeks compared to 1–2 weeks, respectively), it should not be the clinicians' view or singular mindset that mood presentations must take priority. By nature of demonstrated and often significant improvement with ADHD management, the severity of other co-occurring mood disorders often improves as a result. This can also often result in the avoidance of polypharmacy by managing "the most wrong thing" first and then determining if additional medication is warranted for any persistence of comorbid symptoms.

There are obvious exceptions to this approach that warrant addressing comorbid conditions first. Examples would include but are not limited to cardiac or thyroid conditions as well as known presence of bipolar disorder or schizophrenia spectrum disorders as medications most frequently used to treat ADHD can significantly precipitate or exacerbate these conditions. The specific treatment strategies of managing comorbidities and complex cases will be discussed in Chap. 8.

## References

Abramovitch A, Dar R, Mittelman A, Schweiger A. Don't judge a book by its cover: ADHD-like symptoms in obsessive compulsive disorder. J Obsessive Compuls Relat Disord [Internet]. 2013;2(1):53–61. https://doi.org/10.1016/j.jocrd.2012.09.001.

Brown TE. (2013). A new understanding of ADHD in children and adults: Executive function impairments (2nd ed.). Routledge.

Dickerson Mayes S, Calhoun S, Mayes R, Molitoris S. Autism and ADHD: overlapping and discriminating symptoms. Res Autism Spectr Disord [Internet]. 2012;6(1):277–80. https://doi.org/10.1016/j.rasd.2011.05.009.

Fadus M, Ginsburg K, Sobowale K, Halliday-Boykins C, Bryant B, Gray K, et al. Unconscious bias and the diagnosis of disruptive behavior disorders and ADHD in African American and Hispanic youth. Acad Psychiatry [Internet]. 2019;44:1–5. https://doi.org/10.1007/s40596-01127-6.

Gelegen V, Tamam L. Prevalence and clinical correlates of intermittent explosive disorder in Turkish psychiatric outpatients. Compr Psychiatry [Internet]. 2018;83:64–70. https://doi.org/10.1016/j.comppsych.2018.03.003.

Gnanavel S, Sharma P, Kaushal P, Hussain S. Attention deficit hyperactivity disorder and comorbidity: a review of literature. World J Clin Cases [Internet]. 2019;7(17):2420–6. https://doi.org/10.12998/wjcc.v7.i17.2420.

Jaffee S, Hanscombe K, Haworth C, Davis O, Plomin J. Chaotic homes and children's disruptive behavior: a longitudinal cross-lagged twin study. Psychol Sci [Internet]. 2012;23(6):644–50. https://doi.org/10.1177/0956797611431693.

Kennedy RJ, Quinlan DM, Brown TE. Comparison of two measures of working memory impairments in 220 adolescents and adults with ADHD. J Atten Disord [Internet]. 2019;23(14):1838–43.

Kooij JJS, Huss M, Asherson P, Akehurst R, Beusterien K, French A, et al. Distinguishing comorbidity and successful management of adult ADHD. J Atten Disord [Internet]. 2012;16(5):5S–8S. https://doi.org/10.1177/1087054711435361.

Lachambre C, Proteau-Lemieux M, Lapage J-F, Bussieres E-L, Lippe S. Attentional and executive functions in children and adolescents with developmental coordination disorder and the influence of comorbid disorders: a systematic review of the literature. PloS ONE [Internet]. 2021;16(6). https://doi.org/10.1371/journalpone0252043.

Lai M, Kassee C, Besney R, Bonato S, Hull L, Mandy W. Prevalence of co-occurring mental health diagnoses in the autism population: a systematic review and meta-analysis. Lancet Psychiatry [Internet]. 2019;6(10):819–29. https://doi.org/10.1016/52215-0366(19)30289-5.

Lee L, Arunajadai S, Mikl J, Erensen J, Goodman D. The burden of attention-deficit/hyperactivity disorder in adults: a real-world linked data study. Prim Care Companion CNS Disord [Internet]. 2023;25(2):22m03348. https://doi.org/10.4088/PCC.22m03348.

Liu J, Dong F, Lee C, Reyes J, Ivanova M. The application of the adult self-report and the adult behavior checklist form to Chinese adults: syndrome structure, inter-informant agreement, and cultural comparison. Int J Environ Res Public Health [Internet]. 2021;18(12):6352. https://doi.org/10.3390/ijerph18126352.

Marshall P, Hoelzle J, Nikolas M. Diagnosing attention-deficit/hyperactivity disorder (ADHD) in young adults: a qualitative review of the utility of assessment measures and recommendations for improving the diagnostic process. Clin Neuropsychol [Internet]. 2018:165–98. https://doi.org/10.1080/13854046.2019.1696409.

National Academies of Sciences, Engineering, and Medicine. Boat T, Wu J. editors. Mental disorders and disabilities among low-income children [e-book]. 1st ed. Washington, DC. The National Academies Press; 2015. 149–153. https://doi.org/10.17226/21780.

Pearl P, Weiss R, Stein M. Mimics of ADHD, medical and neurological conditions. In: Adler L, Spencer T, Wilens T, editors. Attention deficit hyperactivity disorder in adults and children. Cambridge: Cambridge University Press; 2015. p. 151–8.

Raiker J, Freeman A, Perez-Algorta G, Frazier T, Findling R, Youngstrom E. Accuracy of Achenbach scales in the screening of attention-deficit/hyperactivity disorder in a community mental health clinic. J Am Acad Child Adolesc Psychiatry [Internet]. 2018;56(5):401–9. https://doi.org/10.1016/j.jaac.2017.02.007.

Sadak J. Attention deficit hyperactivity disorder misdiagnosis: why medical evaluation should be a part of ADHD assessment. Brain Sci. 2023;13:1–2. https://doi.org/10.3390/brainsci13111522.

Sahoo M, Biswas H, Padhy S. Psychological co-morbidity in children with specific learning disorders. J Prim Care Spec [Internet]. 2015;4(1):21–5. https://doi.org/10.4103/2249-4863.152243.

Sanches M, Bauer I, Galvez J, Zunta-Soares G, Soares J. The management of cognitive impairment in bipolar disorder: current status and perspectives. Am J Ther [Internet]. 2015;22(6):477–86. https://doi.org/10.1097/MJT.0000000000000120.

Sokolova E, Oerlemans A, Rommelse N, Groot P, Hartman C, Glennon J, et al. Comorbidity between attention deficit hyperactivity disorder (ADHD) and autism spectrum disorder (ASD). J Autism Dev Disord [Internet]. 2017;47:1595–604. https://doi.org/10.1007/s10803-017-3083-7.

Szynabski K, Sapanski L, Conway F. Trauma and ADHD—association or diagnostic confusion? A clinical perspective. J Infant Child Adolesc Psychother [Internet]. 2011;10(1):51–9. https://doi.org/10.1080/15289168.2011.575704.

# Medication Treatment for Children, Adolescents, and Adults with ADHD

7

Ryan J. Kennedy

## 7.1 Disclaimer

The practice of psychopharmacology is as much an art as a science. It is imperative that all treatment decisions remain within accepted standards of care and align with the clinician's scope of practice, education and training, and clinical judgment. While every effort has been made to ensure that the content of this chapter reflects current practices and up-to-date information, the rapid advances in medical research and evolving clinical guidelines mean that some details could become outdated or revised in the future.

The information presented here aims to provide a balance of established guidelines and the author's clinical insights. Readers are strongly encouraged to consult additional authoritative sources (e.g., recent guidelines, pharmacology texts) for a more in-depth exploration of specific topics and to stay current with new developments. Sections of this chapter reflect the author's personal approaches and should not be interpreted as universal practice standards. Practitioners must familiarize themselves with FDA indications, clinical guidelines, and applicable state and federal laws related to ADHD medications. Standard uses of these medications have limitations; before attempting any unconventional dosing or combination therapy, it is advisable to consult or refer to a specialist. When considering augmentation or off-label strategies, thorough research and understanding of the agents are essential.

As with all psychopharmacological treatments, it is critical to monitor key clinical parameters (e.g., blood pressure, heart rate, weight, height) routinely and to be vigilant for potential drug interactions—especially between ADHD medications and any other medications or supplements the patient is taking. Certain complex medication regimens should be managed only by, or in collaboration with, clinicians who have specialized expertise (e.g., co-administering multiple psychotropics, or treating ADHD in the context of significant cardiac issues or other comorbidities).

R. J. Kennedy (✉)
Pediatric & Adult Clinic for ADHD & Related Disorders, Manhattan Beach, CA, USA
e-mail: info@KennedyADHDclinics.com

© The Author(s), under exclusive license to Springer Nature Switzerland AG 2026
R. J. Kennedy (ed.), *Assessment & Treatment of ADHD*,
https://doi.org/10.1007/978-3-032-24171-9_7

The author(s) and publisher disclaim any liability arising from the use or interpretation of the information in this chapter. Ultimately, safe and appropriate prescribing rests on the clinical provider's judgment and the individual patient's circumstances. This content is intended to support, not substitute for, the clinician's own expertise and decision-making process.

## 7.2 Overview of ADHD Medication Treatment

Pharmacologic treatment of ADHD is guided by well-established diagnostic criteria emphasizing persistent inattention and/or hyperactivity-impulsivity with functional impairment across settings (American Psychiatric Association 2013), and is best understood within a lifespan neurodevelopmental framework in which executive dysfunction, impaired response inhibition, and motivational deficits shape both medication selection and treatment goals (Barkley 2014). Medication alone cannot *cure* ADHD, but it can significantly reduce symptoms. Importantly, symptom reduction is a secondary goal—the primary objective is to enhance functioning and overall quality of life. ADHD medications enable a person to function more effectively; however, the extent of improvement varies with each individual's unique response and depends on factors like the drug's potency, duration of action, and practical issues such as cost, availability, and side effects. There is no permanent cure for ADHD, just as hearing aids and eyeglasses do not cure hearing or vision loss—they simply help manage the condition. Similarly, ADHD medication requires an initial trial period and dose adjustments by the provider until an optimal regimen is found; even then, periodic adjustments are often necessary to maintain effectiveness, safety, and tolerability.

An analogy can be made to diabetes management: untreated ADHD (like uncontrolled diabetes or hypertension) carries serious risks. Successful treatment often involves a combination of short- and long-acting agents to cover a person's full daily schedule. For example, diabetes care might use a long-acting insulin for baseline control and additional doses around meals or exercise. In ADHD, the day might begin with a long-acting stimulant to cover work or school hours, with a shorter-acting dose later for homework or evening events. A multimodal treatment approach—including medication plus behavioral or lifestyle adjustments and a multidisciplinary support team—is essential for managing chronic conditions like ADHD and maintaining quality of life.

## 7.3 How Pharmacological Treatments Affect the Brain

The neurobiology of ADHD is complex, but the treatments that have the most robust effects, psychostimulants, are two key brain chemicals: dopamine (DA) and norepinephrine (NE). These neurotransmitters play pivotal roles in regulating attention, effort, motivation, memory, and impulse control—functions often impaired in individuals with ADHD.

Non-stimulant medications (e.g., atomoxetine or guanfacine) work by targeting specific receptors or transporters on neurons that control the release or reuptake of these brain chemicals. For example, atomoxetine blocks the NE transporter, increasing norepinephrine (and indirectly affecting dopamine) levels in the prefrontal cortex, which helps improve concentration and reduce hyperactivity/impulsiveness. In contrast, stimulant medications directly increase the availability of DA and NE: methylphenidate does so by blocking the dopamine and norepinephrine transporters, while amphetamine not only blocks transporters but also releases dopamine from presynaptic vesicles. The alpha-2 agonist non-stimulants (guanfacine, clonidine) work post-synaptically to modulate NE signaling and thereby strengthen prefrontal cortex networks. Amphetamines and methylphenidate differ mechanistically, with amphetamines promoting presynaptic catecholamine release in addition to reuptake inhibition, contributing to their generally greater effect sizes in clinical trials (Faraone 2018).

In practical terms, dopamine helps filter out distractions and supports motivation and task persistence—think of it as enabling you to stay on homework or finish a work project. Norepinephrine helps signal what to focus on and when to act, like giving you the push to start a difficult task or shift gears when needed. When the balance of these chemicals is off (as in ADHD), a person may struggle with procrastination, disorganization, or distractibility. ADHD medications aim to restore the optimal balance of DA and NE, thereby reducing core symptoms and helping individuals better manage day-to-day life.

### 7.3.1 How Stimulants Affect Executive Function

Conceptualizing ADHD as a disorder of the brain's cognitive management system provides a mechanistic rationale for individualized pharmacotherapy that targets specific domains of executive dysfunction rather than relying solely on symptom-based prescribing (Brown and Kennedy 2021). Understanding how neurotransmission impacts *executive functions* can guide medication selection for better outcomes. Key receptors in the prefrontal cortex (PFC) include dopamine D1 receptors, norepinephrine alpha-2A receptors, and others; DA and NE work in tandem in the PFC, with NE enhancing the "signal" (relevant information) and DA reducing "noise" (distractions). This coordination is crucial for executive functions—task initiation, ignoring distractions, sustaining motivation, organizing and planning, shifting focus, and regulating emotion and behavior. In ADHD, these processes are impaired, leading to disorganization, procrastination, and distractibility (Arnsten 2009).

Stimulant medications exert therapeutic effects primarily through inhibition of dopamine and norepinephrine reuptake, enhancing catecholaminergic signaling in prefrontal cortical networks implicated in attention and executive functioning (Stahl 2010). Stimulant medications support PFC neurotransmission by boosting catecholamine signaling, but their effect is dose-dependent. Notably, too little catecholamine activity is ineffective, while too much can impair PFC function (e.g., excessive NE can trigger a stress response that actually worsens attention).

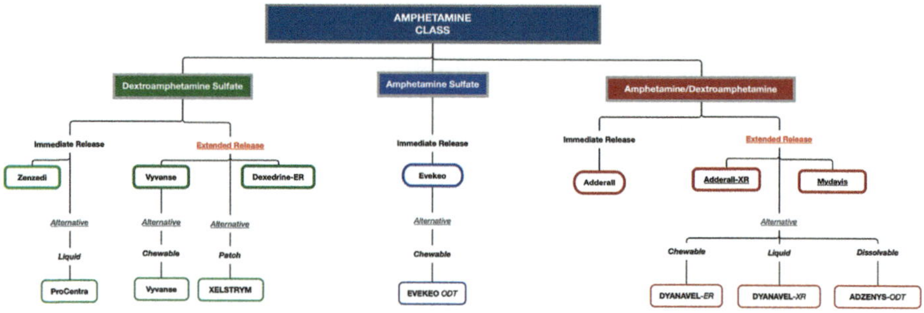


**Fig. 7.1** Amphetamine Class Organized by Compound (Created by Ryan J Kennedy)

This underlines the importance of careful prescribing and encouraging patients to manage stress, as even a correct medication can be undermined by external stressors. Arnsten (2009) describes an "optimal" range of catecholamine activity in the PFC—often depicted as an inverted U-shaped curve—where cognitive function is maximized.

Figure 7.1. illustrates this inverted-U dose–response curve or therapeutic window. There is a sweet spot in which the stimulant dose provides optimal PFC functioning and symptom reduction. Doses below this range are sub-therapeutic (insufficient DA/NE improvement), resulting in little clinical benefit. Doses above the optimal range can flood the PFC with catecholamines and *worsen* functioning, as side effects emerge and the brain's signal-to-noise processing deteriorates. In practice, this means if a dose is too low, the patient may report no improvement; if the dose is slightly too high, the patient may experience new problems (e.g., jitteriness, emotional lability) despite some symptom relief. Titrating to the just-right dose is therefore critical.

Properly fine-tuned stimulant therapy helps patients achieve a therapeutic response that optimizes executive functioning and minimizes ADHD symptoms. However, finding this optimal dose requires individualized titration (discussed in Sect. 7.9.8) and monitoring.

### 7.3.2 How ADHD Medications Are Similar and Differ from Other Psychiatric Treatments

Psychiatric medications, including all of the ADHD medications, cure nothing. "They are not like antibiotics that may cure an infection if the medication is taken consistently for a few days or weeks. ADHD medications are more like eyeglasses that may improve or even normalize a person's vision while the glasses are worn, but the eyeglasses cannot fi x the problem with the person's eyes. When the eyeglasses are taken off, the person's vision returns to whatever it was previously" (Brown 2013). ADHD medication response does not follow the same guidelines as treatments for many other conditions. For example, when titrating levothyroxine for hypothyroidism, we rely on blood levels of thyroid hormone to find a target dose. In ADHD, there is no simple blood test or target level to guide stimulant dosing. Prescribers instead follow FDA-approved dose ranges (primarily to ensure safety and avoid excessive dosing), but these ranges—derived mostly from pediatric clinical trials—have limitations. Some patients may require doses outside the typical range for optimal benefit, whereas others do well below the usual dose threshold. In other words, one should dose to the patient's clinical response rather than strictly to a predetermined target (while staying within safe limits). Flexibility in dosing is reserved for outliers who don't fit the norms, but careful documentation and justification is needed if going above standard recommendations.

When starting a new ADHD medication, begin with the lowest available dose and titrate weekly (or every few days) until the optimal dose is reached without significant side effects. Unlike many medications, do not substitute one amphetamine product for another on a milligram-per-milligram basis—different formulations (even within the same class) have different pharmacokinetic profiles and potencies. In practice, the most effective regimen is often *not* a single daily dose; many patients require a morning dose plus a later dose to maintain coverage into the evening. Despite manufacturer claims that some extended-release stimulants last 8–12 h, however, immediate-release tablets typically last only ~3–4 h, and most generic extended-release capsules last around 6–8 h for many patients. It is common to combine short-acting and long-acting formulations to tailor coverage to a patient's schedule.

When formulating a treatment plan, consider the patient's diagnosis, lab results (if any), medical history, and psychiatric history. Discuss medication options with the patient (and family, when appropriate) so they can contribute their preferences, goals, and report any past medication trials. Shared decision-making in choosing the medication often improves adherence and outcomes.

## 7.4 Classification of ADHD Medications

ADHD medications are commonly classified into two broad categories: **stimulants** (also called *central nervous system stimulants*) and **non-stimulants**. Another way to classify them is by their mechanism of action on the neurobiology of ADHD. For brevity, this chapter uses the terms *stimulants* and *non-stimulants*. However, when educating patients, it may help to use more descriptive (and less stigmatizing) terms. For example, NPs might explain to patients: "Our brain cells communicate using electrical impulses and about 50 different chemicals. ADHD medications primarily affect two of these chemicals, norepinephrine and dopamine. For people with ADHD, the medication doesn't artificially add these chemicals or give you more than normal; rather, it helps the neurons release and reload these chemicals more efficiently, improving communication between brain cells needed for focus, memory, self-control, and learning."

There are over **30 FDA-approved medication preparations** in the United States for treating ADHD across various age groups. These include **seven** distinct active compounds within the two major stimulant classes. The stimulant classes are:

Stimulant Medications AMP: Amphetamine, Dextroamphetamine, Lisdexamfetamine
MPH: Methylphenidate, Dexmethylphenidate, Serdexmethylphenidate

Non-stimulant Medications A2A: Guanfacine ER, Clonidine ER
NRI: Atomoxetine, Viloxazine ER

Emerging pharmacogenetic data suggest that variability in dopamine transporter and CYP2D6 genotypes may influence differential treatment response to methylphenidate and atomoxetine, highlighting the potential for personalized medication selection (Bishop et al. 2024). These findings support a shift away from a purely trial-and-error approach toward a more biologically informed framework, in which individual differences in neurotransmitter systems, metabolic pathways, and receptor sensitivity are considered when selecting and titrating medications. For example, genetic variation in the dopamine transporter may partially explain heterogeneity in stimulant response, while CYP2D6 metabolizer status can influence atomoxetine plasma levels, side effect burden, and overall tolerability. Although pharmacogenetic testing is not yet standard of care in ADHD treatment, these data underscore the importance of integrating patient-specific factors—including genetics, comorbidities, developmental stage, and prior medication response—into clinical decision-making. This individualized approach aligns with broader trends in precision psychiatry, where the goal is to optimize efficacy while minimizing adverse effects through tailored treatment strategies rather than uniform prescribing algorithms. In addition, stimulant formulations can be characterized by their isomeric composition (for instance, some use only the dextro-isomer of the molecule for a more targeted effect) and by their delivery systems. To better understand stimulant isomers (Aiken et al. 2022), authors in Chapter 5 of the book "Prescribing Psychotropics," use the illustration of a pair of scissors to describe left and right isomers. "Isomers are mirror image molecules of one another, and each stimulant comes into isomer forms. Although they look alike, they don't work the same way in the brain, just as a right handed scissors. Don't work well for left-handed people."

There are at least five distinct formulation technologies that allow fine-tuning of how the drug is released and absorbed. For example, stimulant medications come in: immediate-release (IR) tablets and beaded, biphasic capsules that release half or less of the drug for a few hours, and the remaining amount is released hours later to cover late-morning to afternoon. The ratio for this two-phase approach is used in nearly all extended-release stimulants; however, the ratio for IR vs XR is relevant for methylphenidate extended-release formulations, where each long-acting drug has its own unique ratio and different amounts of the drug are delivered.

Each stimulant compound is typically available in a short-acting form (lasting ~3–5 h) and one or more extended-release (XR) formulations. Many older long-acting methylphenidate products use beaded biphasic release technology—combining IR and XR beads in a fixed ratio (e.g., 30% immediate/70% delayed). Newer long-acting options include osmotic-release systems, liquid or chewable capsules containing microbeads, and transdermal patches. This variety allows customization of onset and duration to fit the patient's daily routine and can help mitigate side effects by choosing the right delivery method.

For non-stimulants, there are currently four options. Two NRIs—atomoxetine and viloxazine ER—are FDA-approved for adults with some age-limitations for children, and three types of Alpha 2 Agonists—guanfacine ER (Intuniv) and two types of clonidine ER (Kapvay and Onyda XR)—are FDA-approved for children and adolescents. In practice, clinicians also use guanfacine ER and clonidine ER effectively in adults (off-label) and may combine them with stimulants for an additive effect. (For example, guanfacine ER is often added to stimulant therapy to help with impulsivity or tics.) Research supports the efficacy and safety of these off label uses in adults, but one should always document the rationale and supporting evidence when prescribing off-label.

**Key points:** Stimulants are the *first-line* treatment for ADHD due to their robust efficacy in reducing symptoms. In general, methylphenidate tends to show the best efficacy/tolerability profile for children and adolescents, whereas amphetamines tend to be more effective (and often better tolerated) for adults. Importantly, a positive response to a stimulant does not confirm an ADHD diagnosis, nor does a lack of response exclude it. In clinical practice, it's not uncommon for patients to have an inadequate response to their first stimulant trial (and sometimes the second), especially in younger children who may be more sensitive to side effects. Finding the right medication often requires some trial-and-error within the stimulant class.

## 7.5 Developing a Systematic Approach to ADHD Treatment

In the appendix, there are several useful tools to aid clinicians in their clinical practices, including a Practitioner's Controlled Substance Checklist, Pre-Stimulant Treatment Safety Checklist, Sample Prior Authorization Documentation, and Dose Conversion References.

A standardized approach in clinical practice is to start patients of all ages using the appropriate compound (amphetamine vs methylphenidate), choosing an immediate-release (IR) or extended-release (XR) delivery system, then starting at the lowest dose and systemically increasingly the dose to the desired response, and adjusting time of dose taken to optimize duration of action for daily coverage of activities. Close monitoring is necessary as adjustments are expected during titration, lowering the dose or shortening the duration if problematic side effects emerge or the maximum recommended dosage is reached. An optimal treatment should include a reduction in symptoms; however, the most important goal is improvement in the patient's functioning respective of their developmental abilities.

1. **Selecting the Appropriate Agent and Formulation**—based on age, comorbidities, and daily schedule.
2. **Individualized Dosing and Titration**—to balance symptom control against potential side effects.

3. **Ongoing Monitoring**—using standardized rating scales and patient-reported outcomes to guide adjustments.

Sometimes the most effective way to help our patients is not by adding another medication but by reassessing what they are already taking. Be willing to de-prescribe medications that are not effective or not medically necessary, especially if they may be interfering with ADHD treatment or causing adverse effects. Prescribing stimulants is not always a straightforward decision, and it's normal for clinicians to feel some uncertainty—especially with concerns about side effects or legal/safety liabilities. Engaging in shared decision-making and maintaining open communication with the patient (and/or caregiver) can help mitigate potential risks and address fears related to treatment. For Schedule II controlled substances (like stimulants), and particularly for off-label regimens, documentation is crucial. Always chart your rationale— cite guidelines, research, or expert consensus that support your plan—or note that you consulted with a specialist or colleague about the decision.

When documenting, you might include a statement such as:

"Treatment options within FDA indications, clinical practice guidelines, or expert consensus were discussed with the patient. This treatment plan was reached via shared decision-making, including a discussion of benefits, risks, side effects, the titration process, and the follow-up plan, and the patient (or guardian) provided verbal consent."

Such documentation not only provides a clear record of informed consent and rationale but also demonstrates your due diligence should questions later arise.

### 7.5.1 Process of Treatment

In Figs. 7.1, 7.2, 7.3, 7.4, 7.5, 7.6 and 7.7 is illustrated a simple process for initiating medication management and follow-up workflows and checklists to cover important steps for patients of all ages in any clinical setting.

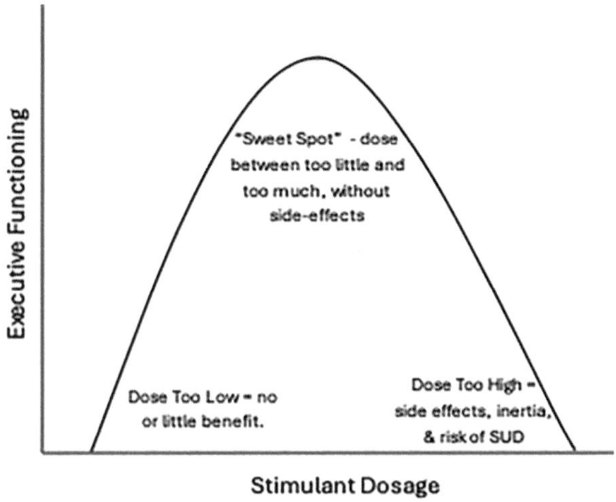


**Fig. 7.2** Inverted U-shaped Dose-Response Curve (Created by Ryan J Kennedy)

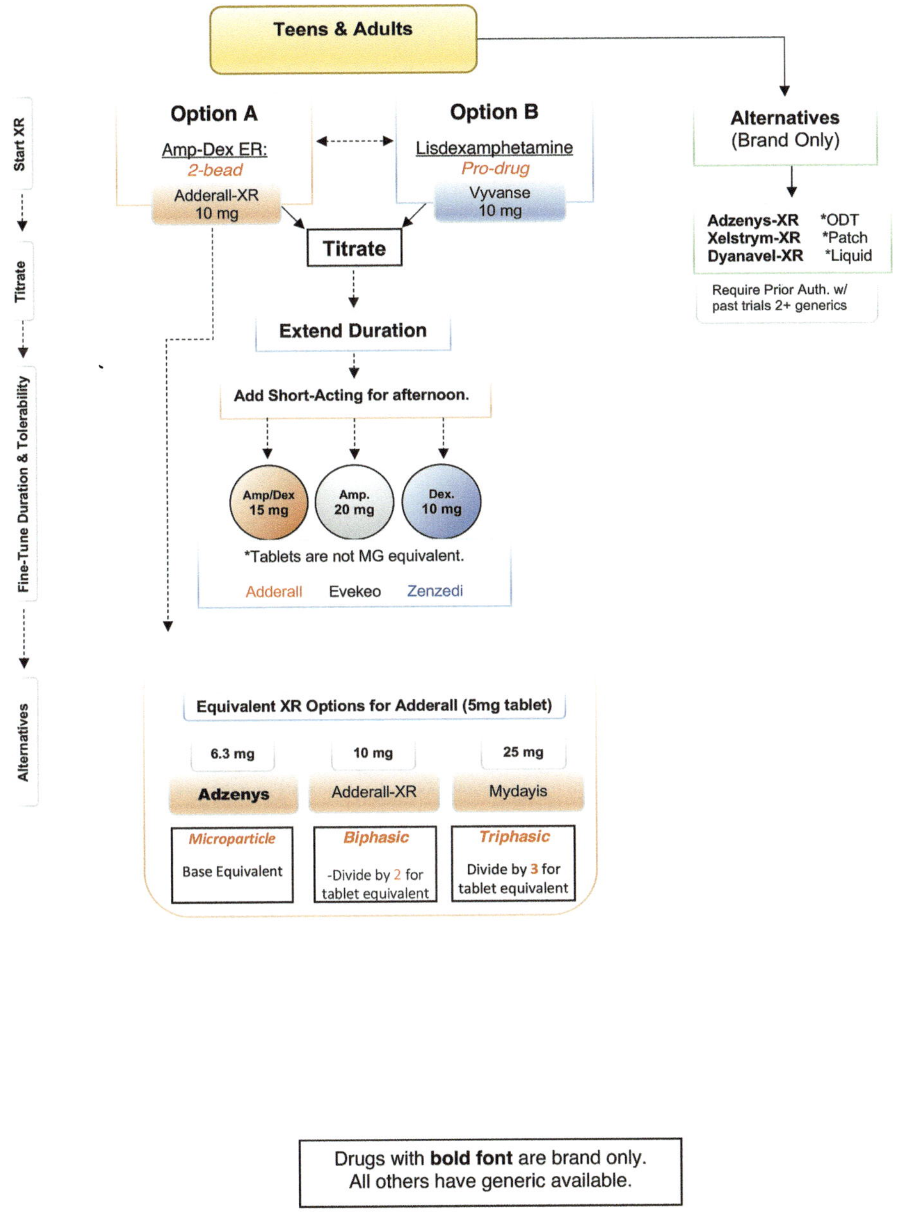


**Fig. 7.3** Starting Amphetamine for Teens and Adults

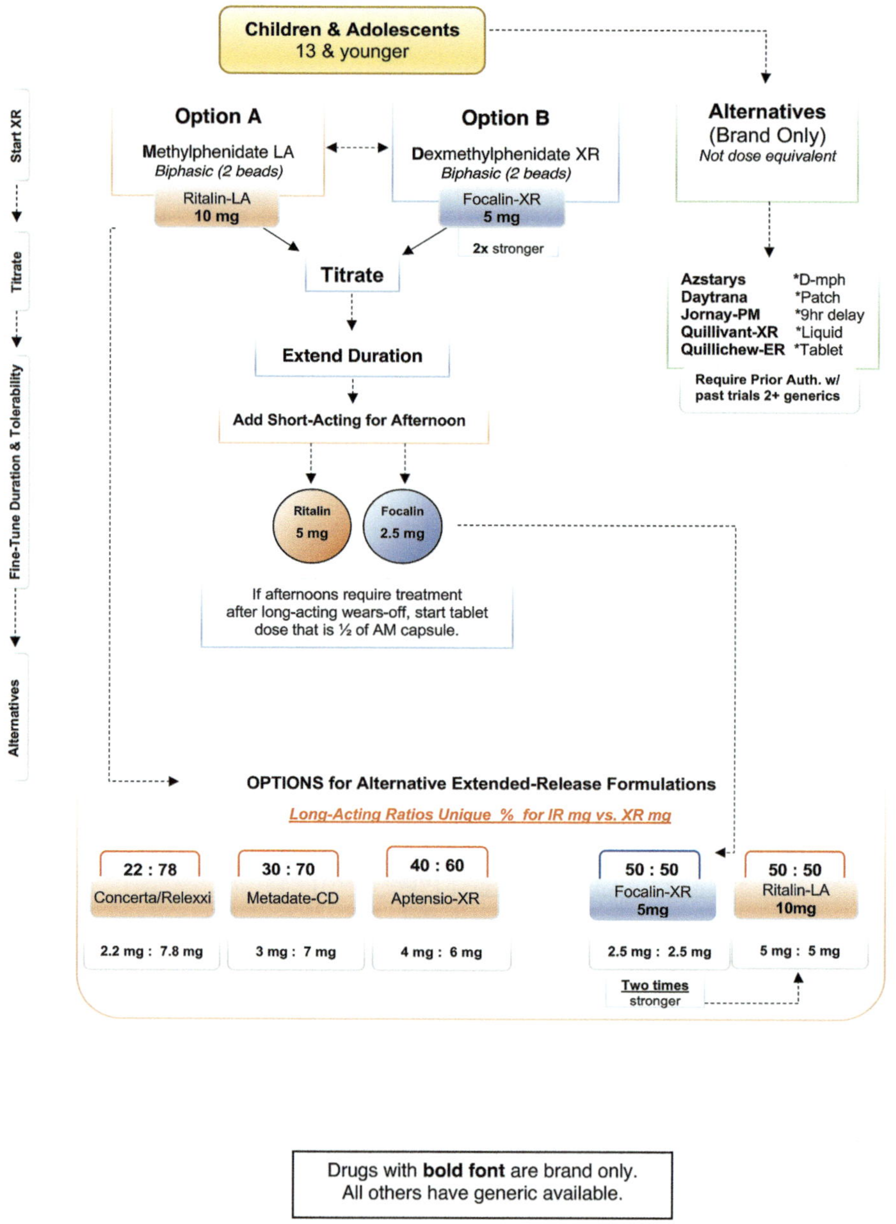


**Fig. 7.4** Starting Methylphenidate for Children and adolescents

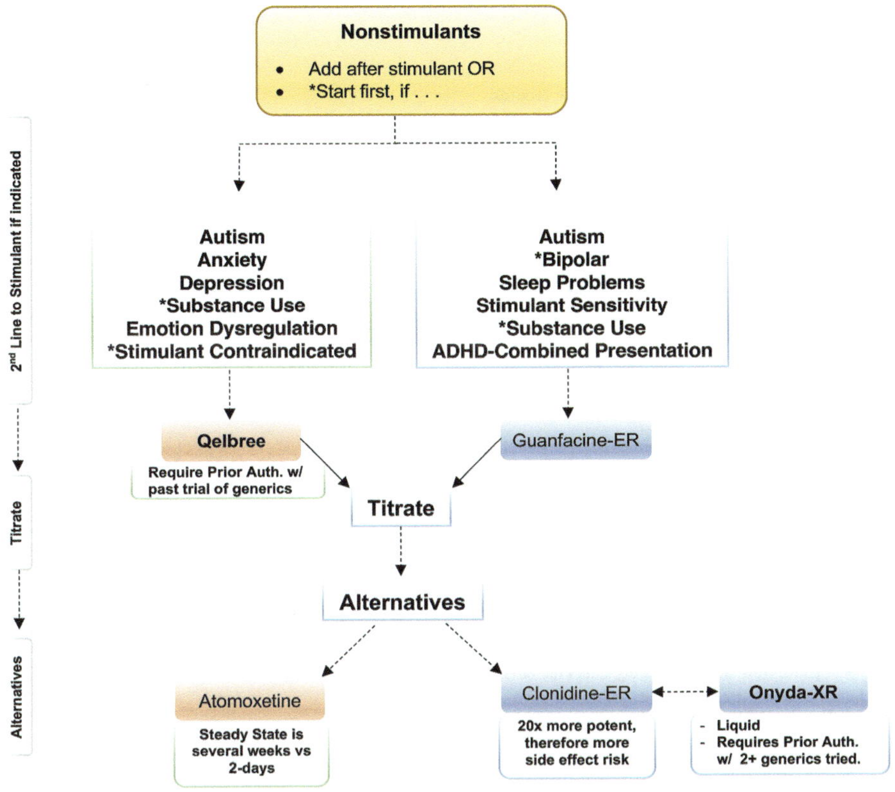


**Fig. 7.5** Starting Treatment with Nonstimulants

It is helpful to have a structured workflow for ADHD medication management that can be used both as an office protocol and a teaching tool when explaining the treatment process to patients and families. Figures 7.1 and 7.2 presents a sample flowchart outlining one approach to initiating treatment and follow-up scheduling for ADHD medication. Key steps in such a workflow include scheduling a follow-up within ~3 weeks of the initial prescription; making dose adjustments at ~monthly intervals during the first 3 months (the "acute titration" phase); fine-tuning the medication over up to ~6 months to optimize the regimen; ensuring an in-person visit during the first month of treatment; establishing a regular follow-up cadence (e.g. telehealth every 3 months, in-person every 6–12 months) with additional check-ins as needed; setting clear policies for patient communication (e.g. reporting response via portal between visits, requiring appointments every 90 days for Schedule II refills); monitoring side effects, sleep, appetite, vitals, growth, and any required lab tests (such as an EKG if indicated); and managing prescriptions (e.g., providing up to three 30-day prescriptions for stable regimens, coordinating with pharmacies for any "hold" prescriptions, and allowing 48–72 h for refill processing). By following a consistent workflow (and adapting it to your setting), you can improve efficiency and ensure important safety checks and patient education steps are not overlooked.

**Initial ADHD Treatment Appointment**

**1. Pre-Visit Preparation**
- ☐ Review referral reason and past records
- ☐ PT/others complete screener/rating scale (Vanderbilt, ASRS...)
- ☐ Check State Controlled Substances Monitoring Program
- ☐ Check formulary for insurance coverage and prior authorization needs

**2. Clinical Assessment**
- ☐ Obtain comprehensive developmental, medical, psychiatric, and family history
- ☐ Assess school performance and behavioral observations
- ☐ Conduct physical exam: vitals, height/weight, BP, HR
- ☐ Assess PT for possible contraindications: severe tics, seizures, cardiac concerns, bipolar, substance use…

**3. Administer Objective Rating Scale Tools for Baseline Data & Goals of Treatment**
- ☐ Executive Function Scale: BEFARS, BRIEF, or CAARS
- ☐ Send teacher and parent versions (if not completed prior)

**4. Diagnostic Impression**
- ☐ Confirm DSM-5 criteria for ADHD subtype
- ☐ Screen for comorbidities (e.g., anxiety, depression, ODD)

**5. Education & Treatment Planning**
- ☐ Discuss diagnosis and treatment goals with family
- ☐ Provide written education on ADHD and treatment options
- ☐ Discuss medication plan: type, **dose range**, **onset**, **duration**, **side effects**, titration schedule
- ☐ Review risks, benefits, and expectations of treatment
- ☐ Review stimulant safety checklist & (if needed) discuss/have patient sign a non-diversion agreement

**6. Medication Initiation**
- ☐ Determine if cardiac workup or EKG is needed
- ☐ If starting stimulant:
  - ☐ Prescribe **1st Rx**: 30-day supply
  - ☐ Plan **follow-up within 3 weeks** (before Rx runs out)
  - ☐ Schedule future visit before end of encounter

**7. Prescribing Instructions**
- ☐ For maintenance plan:
  - Provide a single 30-day prescription
- ☐ Use EHR messaging to encourage reporting of side effects

**8. Documentation**
- ☐ Note intended **treatment goals** (e.g., focus, impulsivity, school performance)
- ☐ Document patient/family understanding and consent
- ☐ Log ADHD rating scale results for baseline

**Fig. 7.6** Initial ADHD treatment appointment

**Follow-Up ADHD Appointment Checklist**

**1. Appointment Interval**

- ☐ For new/adjusted dose: schedule follow-up **within 3 weeks**
- ☐ For maintenance: schedule every **90 days minimum**
    - Schedule next visit at **3rd week of 3rd month** from today

**2. Clinical Assessment**

☐ Before Appointment, have patient complete same rating scale
☐ Review response to medication:

- ☐ Onset of effect: ______
- ☐ Duration: ______ # of hours
- ☐ Any trough symptoms
- ☐ Rebound effects

- ☐ Assess time of dose, onset, duration; adherence (weekdays vs weekend), any missed doses & response.
- ☐ Vitals: HR, BP, weight, height
- ☐ Use ADHD rating scale (e.g., follow-up Vanderbilt, ASRS)

**3. Side Effects & Safety**

- ☐ Appetite, sleep, mood, tics, headaches, GI symptoms
- ☐Monitor for side effects, misuse, tolerance, or abuse
- ☐ Check PDMP for monitoring controlled substances prescribed

**4. Medication Plan**

- ☐ Adjust dose/formulation if needed & discuss target/optimal dose goal
- ☐ If stable: issue 3 monthly Rx's with format below:
    - "Rx 1 of 3" → Fill 28–30 days after last Rx or [insert exact date]
    - "Rx 2 of 3" and "Rx 3 of 3" similarly labeled as above; ** *see Practice Tip and date table below.*
- ☐ Update goals based on function/symptom change

**5. Communication Plan**

- ☐ Remind patient how to share feedback, report side effects... EHR portal, phone, secure text
- ☐ Reinforce NOT to change dose without contacting provider

**6. Documentation**

- ☐ Objective ADHD scale comparison (baseline vs now)
- ☐ Note functioning in daily life
- ☐ Record medication response and next planned step

**** Clinical Practice Tip for Charting & Writing Prescription Dates**

- Use this practice tip to copy/paste dates faster for typing in the dates needed before sending multiple prescriptions.
- This format makes it easy to copy/paste prescription dates directly into the EHR and ensures consistent scheduling of follow-up appointments and medication management
- Create a 4-column chart in Google Sheets.
- Label the columns and rows with the example format in table below.
- Add more rows for everyday of year by dragging down to autofill future dates.

**Fig. 7.7** Follow-up ADHD appointment checklist

## 7.6 Evidence Base for ADHD Medications

The volume of information on ADHD pharmacotherapy—from textbooks and guidelines to research studies and media reports—can be overwhelming. Clinicians must critically appraise the source and quality of evidence before applying recommendations to patient care. Here we distill some key evidence-based guidance.

### 7.6.1 Limitations of FDA: Dosing Limitations

Given the lack of data in adult ADHD clinical trials and nature of prescribing Schedule II psychostimulants, there are fears and doubts about the maximum dose range. Some clinicians may interpret 40 mg of amphetamine to be the maximum effective dose, and others may only go up to 60 mg in severe cases. It is important to remember that the optimal dose for any individual is not based on weight and severity of symptoms nor outdated clinical trial data. However, limitations exist for an important reason, and higher dosages do come with greater risks and concerns with psychological or chemical dependency. In clinical practice, if an adult patient is determined to have the maximum benefit without side effects using lisdexamfetamine 50 mg once daily in the morning, and during follow-up, the NP inquires about duration of action and finds that the this effective dose actually only lasts 6 h.

Tables 7.1 and 7.2 provide the actual base equivalency for two common medications used in ADHD treatment. This approach provides a more scientific baseline for amphetamine dosage and would assist in more precise delivery of care.

**Table 7.1** Total Amphetamine Base Equivalence of Adderall (note that base equivalence is standard doses of Adzenys-XR)

| Amp/Dex tablets | Base equivalency (actual mg) |
|---|---|
| 5 mg | 3.13 mg |
| 7.5 mg | 4.7 mg |
| 10 mg | 6.3 mg |
| 12.5 mg | 7.8 mg |
| 15 mg | 9.4 mg |
| 20 mg | 12.6 mg |
| 30 mg | 18.8 mg |

Adderall and Adzenys-XR ODT (U.S. Food and Drug Administration 2025a, 2025b)

**Table 7.2** Total Amphetamine Base Equivalence of Vyvanse

| Lisdexamfetamine capsules | Base equivalency (actual mg) |
|---|---|
| 10 mg | 5.8 |
| 20 mg | 11.6 |
| 30 mg | 17.3 |
| 40 mg | 21.1 |
| 50 mg | 28.9 |
| 60 mg | 34.7 |
| 70 mg | 40.5 |

Vyvanse (U.S. Food and Drug Administration 2025c)

### 7.6.2 FDA Limitations for Drug Labels and Age Restrictions

Clinicians often encounter situations where the FDA-approved age indication of an ADHD medication doesn't align with the patient's needs. A common tip-off is a prior authorization request from insurance due to the patient's age. For example, an adult patient prescribed guanfacine ER (Intuniv) will likely trigger a denial because this medication's label is for age 6–17. Yet guanfacine can be effective in adults, and in fact it *has* been studied in adults (two controlled trials in Japan included US adults). Similarly, we face paradoxes like dextroamphetamine IR tablets: they are FDA-approved for ages 3–17, but in practice dextroamphetamine is usually considered a *third-line* choice for young children (after trying MPH and perhaps a non-stimulant) due to side effect concerns. Meanwhile, Adderall IR (mixed amphetamine salts) is widely prescribed to adults even though the FDA never officially approved the immediate-release form for adults (only the XR form gained adult approval, in 1996). The take-home point is that strict age criteria in FDA labels should not be the sole determinant of medication choice. Clinical evidence and patient specific factors should prevail. If you have a sound reason to use a medication off-label for a given age group (backed by literature or guidelines), it's appropriate to do so—with documentation. For instance, Meta-analytic findings in adults indicate that both standard and higher-than-licensed stimulant doses may be associated with improved symptom outcomes, although this must be balanced against tolerability and safety considerations (Farhat et al. 2024). Furthermore, an excellent source of clinical information can be found in the paper by Cortese et al. (2018) who published a comprehensive network meta-analysis that cites an evidence-based resource for selecting the most efficacious, safe, and well-tolerated medications by age group. It is useful to reference such evidence when discussing initial medication selection with patients, documentation in charting, and use for submission or appealing a prior authorization for a particular class or drug.

### 7.6.3 How to Interpret Research Statistics

When evaluating ADHD medication studies—especially meta-analyses—clinicians should look beyond p-values and consider *effect size*, which quantify the magnitude of a treatment's benefit. The *standardized mean difference (SMD)* is a common effect size metric that allows comparison of drug effects across trials with different outcome scales. Conventionally, to interpret SMD values as follows (roughly analogous to Cohen's *d*) (Table 7.3):

**Table 7.3** How to interpret effect sizes with standardized mean difference (SMD)

| Effect size magnitude | SMD range |
|---|---|
| Small | < 0.3 |
| Medium | < 0.7 |
| Large | 0.8 or > |

**Table 7.4** Comparison of effect size (SMD) of ADHD medications in children and adults

| Medication | Children | Adults |
|---|---|---|
| Methylphenidate | 0.78[a] | 0.49[a] |
| Amphetamines | 0.95[a] | 0.79[a] |
| Atomoxetine | 0.32[a] | 0.45[a] |
| Guanfacine ER | 0.67[b] | 0.52e |
| Viloxazine | 0.47[c] | No data |
| Clonidine ER | 0.30[d] | No data |

Sources:
[a]Cortese et al. (2018)
[b]Lenzi et al. (2018)
[c]Moukhtarian et al. (2017)
[d]Connor et al. (2010) e Iwanami et al. (2020)

Generally, the higher the SMD score indicates a greater symptom reduction; however, this measure is not intended to guarantee an individual's probability of an optimal medication for treatment outcome, *nor does it account for tolerability*. For example, amphetamines in children often achieve SMDs >0.7 (a large effect), and methylphenidate yields a similar effect size on average. However, amphetamines tend to cause more side effects in children, so the "best" choice isn't based on efficacy alone. On average, pooled stimulant trials show much higher effect sizes (~0.9) than non-stimulant trials (~0.45), but that doesn't mean non-stimulants have no place—individual tolerability and comorbidities might make a non-stimulant the right choice for a particular patient.

To aid in treatment selection, Table 7.4 summarize pooled effect sizes from double-blind, randomized controlled trials or synthesized in network meta-analyses or large systematic reviews for the major medication classes in children and in adults. Use these data as a general guide when comparing agents, but always weigh effect size alongside each patient's unique situation (tolerability, comorbid conditions, preference, etc.). For brevity and clinical utility, Table 7.4 lists one representative effect size per medication for adults and children; however, network meta-analyses typically report a range of SMD values depending on outcome measure, informant/rater, study duration, dose, and analytic model. As shown in Table 7.4, although amphetamines demonstrate the largest effect sizes, this does not make them the preferred first-line option in children given their comparatively greater side effect burden and tolerability concerns which is not factored into the effect size (SMD); methylphenidate is often the more appropriate initial choice. Similarly, lower effect sizes observed with non-stimulants should not be interpreted as inferiority, as these agents are generally better tolerated, and their use—particularly in combination with a stimulant—may provide improved overall symptom control than monotherapy.

By anchoring clinical decisions in both effect size *and* individual patient factors, NPs can translate statistical data into personalized, evidence-based ADHD treatment plans.

### 7.6.4 Generic vs Brand-Name Considerations

When a stimulant medication is available in both brand and generic versions, it's important to understand what "bioequivalence" means. To be considered bioequivalent, the FDA requires a generic's maximum concentration **(C-max)** and area under the curve **(AUC)** to be within an acceptable range compared to the brand product. These measures ensure similar overall exposure to the drug. However, for ADHD stimulants, the perceived duration of effect is most often related to the range of time surrounding its peak concentration **(T-max)** and the release mechanism, than max concentration (C-max). In practice, some patients report that a particular generic "doesn't last as long" or "kicks in differently" than the brand, even if blood levels are considered equivalent. Minor differences in the formulation can affect the release profile.

Also, note the distinction between **authorized generics** and regular generics. An authorized generic is the brand-name product packaged as a generic (identical ingredients and formulation), whereas a typical generic is a different formulation that has demonstrated bioequivalence. Authorized generics are essentially the same as brand, so you wouldn't expect any difference in performance. With other generics, most patients do fine, but a subset may respond to one version slightly differently than another. It can be helpful to ask patients if they noticed a change when their pharmacy switched manufacturers.

### 7.6.5 Navigating Insurance Formularies

Health insurance and their inconsistent formularies for ADHD medications are a major issue and a significant barrier to treatment selection. Newer and brand-name medication typically require prior authorizations and typically require trials of at least 2–3 generic or tier 1 or 2 medications that are in the same class or equivalent to the request brand name drug. This process is burdensome and can be a significant barrier to treatment adherence and follow-up appointments for fine-tuning medications. For instance, a patient might really benefit from Vyvanse, but many insurers demand that generic Adderall IR and XR be tried (and failed) first due to cost. Navigating these requirements is unfortunately part of practice—often it means documenting those trials or writing a justification letter. There are templates in the **Appendix** which can be modified for use in clinical practice when insurance requests a prior authorization and supporting documentation.

## 7.7 Developing a Process of Treatment

### 7.7.1 Selection Based on Age

**Pediatric and Adolescent Patients** The Society for Developmental and Behavioral Pediatrics (SDBP 2019) and the American Academy of Pediatrics (2019) both endorse **methylphenidate** as first-line therapy for school-aged children with ADHD. (Amphetamines may be used, but MPH has a slight edge in tolerability for younger patients.)

**Adolescent and Adult Patients** Expert consensus (e.g., the World Federation of ADHD (Faraone et al. 2021) and available adult guidelines recommend **amphetamine**-based stimulants as the preferred initial treatment for adolescents and adults. Notably, the American Professional Society of ADHD and Related Disorders is expected to publish the first dedicated adult ADHD guidelines in the USA in 2026.

### 7.7.2 Selecting a Delivery System

**Tablets and Capsules** When prescribing, give clear instructions about how to take the medication. For instance, extended-release capsules or tablets should be swallowed whole and not crushed or chewed (crushing an XR capsule can release the full dose at once, increasing side effects). Some capsules can be opened and sprinkled on food if a child can't swallow pills—check which ones are designed for that. Explain that different XR products have different release profiles (e.g., Concerta's OROS vs Adderall XR's beaded technology), but patients shouldn't worry about those specifics beyond taking it as directed.

**Alternatives** If a patient has difficulty swallowing pills, there are ways to manage this. Practice in the office with small candies (like using mini M&M's, then regular M&M's, as pill-size analogs) can help overcome the psychological hurdle. If that fails, remember there are alternative formulations: Quillivant XR (methylphenidate) is a long-acting liquid; Cotempla XR is an oral dissolving tablet; and Daytrana and Xelstrym are transdermal patches for methylphenidate and amphetamine). Adzenys-XR ODT, Vyvanse, Dyanavel, and QuilliChew ER are chewable or orally disintegrating long-acting formulations of amphetamines and methylphenidate. These can be especially helpful for kids who cannot swallow capsules. Many people do not realize how much variability there is in individual responses to stimulants. Two patients of the same age and size can have very different reactions to the same dose of medication. Interestingly, an individual's response to stimulants does not depend on their age, gender, height, weight, or even baseline severity of symptoms. Instead, it is largely determined by how their body metabolizes the drug and their personal sensitivity to it. A very knowledgeable clinician with a carefully designed regimen can still encounter common pitfalls: for example, a dose that doesn't last long enough, a dose that's too high, or a regimen that was working well but then became less effective. The sections below discuss strategies to address these scenarios.

### 7.7.3 Strategies for Optimization

**Dose Timing and Frequency** Understanding the patient's daily schedule and how many hours of symptom coverage they need is crucial when selecting a medication and dosing schedule. Many children, teens, and adults do well taking a long-acting stimulant once in the morning, which provides coverage for a typical school or work day. If the medication consistently wears off at, say, 4 PM, an afternoon booster dose can be added to cover the late afternoon and evening. Patients who need to focus in the evening (for homework or a second job, for example) might use a short-acting stimulant in the early evening so that it doesn't interfere with sleep. In practice, you will tailor these decisions: someone with an early morning job might even need a predawn short-acting dose to get going, whereas someone who only struggles in the late afternoon might only need a small 4 PM dose.

**Extending the Duration of Action with Combination of Short and Long-Acting Stimulants** Individual patients report differences in duration of action between certain generic and brand formulations, possibly due to those release profile nuances. Clinicians should be aware of this, and, if a patient experiences issues with a generic, consider whether trying a different manufacturer or the brand version is warranted.

**Case Example** Patient is a 19-year-old college student with ADHD. His class and work schedule vary: on some days he works 8:00 AM to 4:30 PM then has evening classes from 7:00 to 11:00 PM; on other days he has classes at 9:00–11:00 AM, 1:00–3:00 PM, and 5:00–7:30 PM. After an initial trial on a single morning long-acting stimulant left him without coverage for evening study, Jon and his prescriber collaborated to increase his regimen to two long-acting doses (morning and mid-afternoon) on work days and one long-acting plus a short-acting dose on heavy class days. This way, he gets through work or daytime classes and can still focus in the evening, without medication effects interfering with sleep. Each semester, Jon brings his new class schedule to his provider, and together they adjust the timing and dosing of his medication regimen for the upcoming term.

The total amount of stimulant medication a patient like Jon takes is unique to his body's needs. Another patient might find that same regimen intolerable. In fact, even a standard once-daily dose can cause some patients to discontinue treatment if side effects are troublesome. This is why it's so important to ask patients about each side effect they experience and when it occurs. Sometimes, a side effect means the dose is too high or that a particular medication is too potent for them. Often these issues can be improved by reducing the dose, adding a booster dose at a different time, or switching to a different medication.

One common scenario is a patient for whom a long-acting stimulant "doesn't last long enough." Although some extended-release stimulants are marketed as lasting a full day, many patients (especially those who metabolize stimulants quickly) may get only 4–6 h of useful effect. This situation is analogous to how some people feel hungry sooner after a meal or require more alcohol to feel an

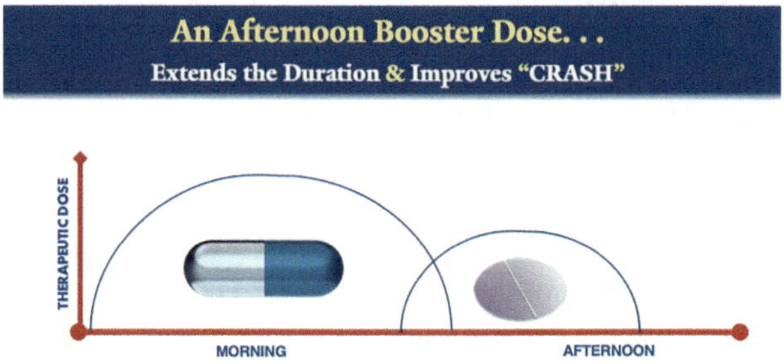


**Fig. 7.8** Combining long and short-acting stimlants. (Created by Ryan J Kennedy)

effect—metabolic differences play a big role. For those who are fast metabolizers of stimulants, a single dose often isn't sufficient. They may request an additional dose, but prescribers are sometimes reluctant to prescribe a long-acting stimulant *twice daily* due to fears of overmedicating, potential for diversion, or simply because it's uncommon.

For example, some patients may feel too "jittery," "tense," "wired," or "irritable" after a stimulant's onset. If this regularly occurs this usually indicates the dose is **too high**—similar to how someone might feel after drinking too much strong coffee or caffeine. The solution is to wait a few more days to allow time to adjust to the dose or simply reduce the dose to the next lowest amount until side effects are manageable, or to try a different stimulant class if even low doses cause this reaction (Fig. 7.8).

Other patients may feel a late-day *crash* of fatigue, irritability, or dysphoria when the stimulant wears off too abruptly. This "rebound" effect can happen soon after the drug's half-life is reached. It can often be smoothed out by a short-acting booster dose in the afternoon, which tapers the drug level more gradually and prevents a sharp drop-off.

As with any medication, patients should consult with their prescriber before making any changes to their stimulant regimen—including skipping doses, stopping medication, or timing adjustments. Emphasize this in your patient education: we can usually fix side effect issues by tweaking the regimen, but we need to be informed in order to help.

**Sample Conversation for Titration Plan** Here is a sample titration plan for methylphenidate using dialogue as if this conversation occurred with. a parent and their child in an office setting.

*"Your child didn't show much improvement on the very low starting dose of Ritalin, which is perfectly normal. The good news is he tolerated it well, so we can proceed to carefully increase the dose in small steps (5 mg at a time) until we see benefits. We'll stick with Ritalin and try to optimize it before considering other medications.*

*Here's how we'll do it: We'll start with 5 mg of Ritalin (which for practical reasons is 2 tablets, totaling 10 mg) in the morning, and 5 mg (2 tablets) around noon. If he doesn't respond to the 10 mg morning dose after a day or two, on the next day we'll increase the morning dose to 15 mg (3 tablets). If that higher morning dose seems like too much for him (if he gets jittery or upset), we'll keep the afternoon dose at 10 mg; but if he handles 15 mg fine, we'll also give 15 mg in the early afternoon.*

*We'll continue this 'stair-step' process—gradually increasing to the next level—until we either see the desired improvement or we reach the maximum dose he can tolerate. Generally, the maximum per dose for Ritalin is about 30 mg (which is 6 tablets at once), and we usually don't go above a total of 60 mg in a day. If at any point you notice that a certain dose seems like too much for him (for example, if 15 mg made him too emotional or flat), we'll go back down to the prior lower dose. Whatever dose ends up being effective and tolerated can later be switched to a long-acting capsule form for convenience."*

*The above is an example of how a provider might communicate a titration plan. It illustrates the stepwise increase and the logic of finding the optimal dose. In practice, be sure to tailor the explanation to the caregiver's/patient's understanding.*

Keep in mind a few guiding principles during titration:

The **maximum recommended dose** of methylphenidate IR is typically 60 mg/day (20 mg per dose, up to 3× daily) for school-age children, though some may need less and a few may benefit from more under close supervision. For amphetamines, usual pediatric max is around 40 mg/day for Adderall XR (though adults may go higher). Always check current guidelines for max doses by age and formulation:

- If a patient reaches a dose that is clearly too high (e.g., significant side effects with no added benefit), step back to the last tolerable dose. That dose can be considered the effective dose, or you might transition that dose to an equivalent extended-release formulation for convenience.
- Communicate frequently during titration. Often, we have patients report in (by phone or portal message) after a week at each new dose. If problems arise, adjustments can be made promptly without waiting for the next face-to-face visit (Fig. 7.9).

For patients who can tolerate max dose of two smaller XR capsules but need a shorter duration than Mydayis or Concerta, Xelstrym. First, consideration must require following a careful documented history that would be supported by scientific rationale, guidance from an expert and document with supporting evidence to clearly explain the rationale and purpose to legitimize medical necessity and statement that consent was provided in the case of liability for potential risks. The potential benefit for patients who are outliers in rapid metabolism should be trusted with the treatment process so long as a structured evaluation and shared-decision making process is followed in a standardized manner.

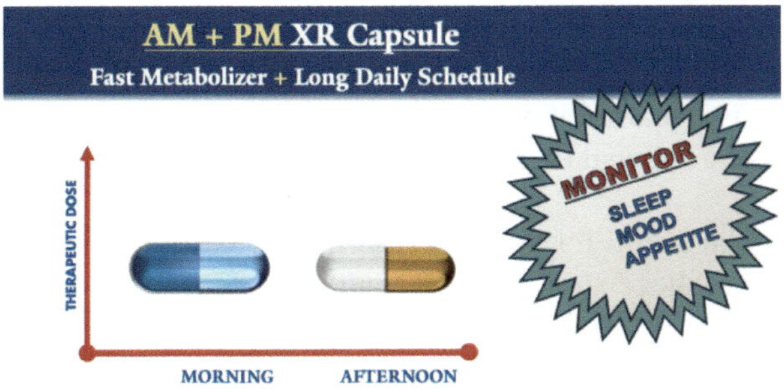


**Fig. 7.9** Long-acting XR twice daily. (Created by Ryan J Kennedy)

More often, clinicians will add a short-acting booster in the afternoon. However, for some patients a more effective strategy is actually two doses of a long-acting stimulant (e.g. one in the morning and one in early afternoon). Using two long-acting doses can provide smoother coverage across an extended day (e.g. an early work shift and evening classes) and may reduce the peaks and troughs that come with an IR booster. Of course, this approach should only be used if the benefits outweigh the risks and side effects, and with careful monitoring.

### 7.7.4 Unique Delivery Systems

For some patients, new **extended-release products** can offer unique benefits. Methylphenidate as a long-acting drug is formulated using unique delivery systems such as Concerta's osmotic-release oral systems (OROS), Quillivant's liquid solution and QuilliChew's chewable tablets that are formulations with microparticles, and Daytrana's transdermal patch. This variety of delivery mechanisms enables clinicians to tailor treatment to each patient's needs, offering different onsets of action, durations, and convenience factors.

**Jornay PM (Methylphenidate at Bedtime)** This is a unique formulation taken at **night**; it has a delayed release (~10 h after ingestion). It's designed for patients who struggle significantly with the morning routine (e.g., getting up, dressing, and out the door). Taken around 8–9 PM, Jornay PM begins to work by the next morning, providing about 8 h of coverage through the first part of the day. This can help "prime" a patient for the day without needing a morning dose. One must titrate Jornay PM slowly (start at the lowest dose and increase weekly) and be mindful that a small percentage of patients might have trouble falling asleep with it (since a tiny portion is released overnight).

**Extended-Release Options in Adolescents/Adults** Older teens and adults usually prefer once-daily dosing if possible. Adderall XR and Vyvanse are two very common choices to start an adult on, each with pros/cons. An even longer-acting option, Mydayis, is essentially an "extended-extended" release mixed amphetamine salts (it's like Adderall XR but with an additional third release phase, lasting up to 16 h). Mydayis is a 3-bead system and Adderall XR is a two-bead system; respectively divide the dosage by 3 and 2 to find the equivalent Adderall dosage. For example, Mydayis 50mg is equal to about Adderall 15mg tablets taken 3 times daily, and Adderall XR 30mg is equal to Adderall 15mg tablets taken 2 times daily.

Mydayis can be useful for an adult who truly needs a medication that lasts into the late evening (college students with long days, for instance) and can tolerate a stimulant that long. Adderall XR, in contrast, might cover a work day but often requires a small IR booster for evening tasks. Note that Adderall XR's FDA max is

60 mg for adults, but most patients don't require that much; Mydayis maxes at 50 mg (due to its longer action). For insurance, Mydayis can be harder to get covered (newer and brand-only), whereas generic Adderall XR is widely covered.

### 7.7.5 How to Determine If Your ADHD Stimulant Dose Is Incorrect

Typically, short-acting ADHD medications begin to take effect within 30–60 min, while longer-acting medications remain active for a few hours, 4–6 h, or even a full day. However, most patients notice that the medication's effects gradually diminish as the afternoon or early evening approaches.

If you experience any of the following symptoms after taking the medication at a time when you anticipate its onset, it may indicate an incorrect dosage or a lack of the appropriate medication:

1. Excessive Energy: Feeling overly wired or jittery, similar to excessive caffeine consumption
2. Irritability: Experiencing heightened irritability, with even minor stimuli causing discomfort
3. Fatigue: Feeling excessively tired or experiencing a sense of heaviness, even with adequate focus

However, if you experience any of these symptoms at a time when you anticipate the medication's effects to diminish and then experience improvement after a few hours, it may indicate a different issue. This phenomenon is often referred to as "rebound symptoms" or the afternoon "crash." In such cases, a small, fast-acting dose of the same medication can be added about an hour before the "crash" would be expected and this facilitates a smoother metabolism of the medication, effectively mitigating the crash symptoms.

### 7.7.6 Pediatric Stimulant Treatment

Bringing the above concepts together, here is a **systematic medication trial algorithm for children (approx. Age < 13)**:

1. **Start with Methylphenidate (Step 1):** Begin with low-dose generic **methylphenidate IR** (e.g., 5 mg twice daily) for at least 1–2 weeks. If some benefit is seen and it's tolerated, you can convert to an equivalent long-acting MPH (e.g., 5 mg BID Ritalin LA 10 mg QAM). Adjust timing or add an IR booster in afternoon as needed for coverage. If MPH works but not long enough, options include moving to a longer-acting MPH (Concerta, Adhansia) or using *two* doses of an intermediate/long-acting MPH (e.g., one before school, one at lunch) to avoid afternoon crash.

2. Or try **dexmethylphenidate (Focalin)**. There are only two longer-acting stimulants for Focalin, which is more common to prescribe long-acting methylphenidate products. It's essential to know the key difference is that dexmethylphenidate is roughly twice as potent as racemic MPH, start at about half the equivalent MPH dose (e.g., if patient got up to 20 mg BID of MPH, start Focalin 5–10 mg BID). Titrate Focalin IR by 2.5–5 mg increments weekly. If effective, consider Focalin XR either once daily or BID. (Using two XR doses—morning and early afternoon—often provides steadier coverage than XR + IR later.) Monitor for less rebound with this approach.

## Methylphenidate

| Drug name | IR vs ER ratio | Time of onset | Peak concentration | Duration of action |
|---|---|---|---|---|
| Focalin XR | 50/50 | 30 min | 2–4 h | 5–7 h |
| Ritalin LA | 50/50 | 30 min | 2–3 h | 5–7 h |
| Aptensio XR | 40/60 | 60 min | 2 and 8 h | 5–7 h |
| Azstarys | 30/70 | 30 min | 3–5 h | 6–8 h |
| Metadate CD | 30/70 | 60 min | 3–4 h | 5–7 h |
| QuilliChew ER | 30/70 | 30 min | 3–4 h | 5–7 h |
| Cotempla XR-ODT | 25/75 | 60 min | 4–6 h | 6–8 h |
| Concerta | 22/78 | 60 min | 4–6 h | 6–10 h |
| Quillivant XR | 20/80 | 45 min | 4–6 h | 6–9 h |
| Jornay PM | 5/95 | 9 h | 14 h | 8–10 h |
| Daytrana | n/a | 2 h | n/a | +2 h after taken off |

In the table below is a detailed example of ratio difference for Aptensio XR's biphasic release of 40% of dose in the immediate-release phase and the remaining 60% of the dose in the extended-release phase.

## Aptensio XR

| Dosage | IR (40%) | (ER) 60% |
|---|---|---|
| 10 | 4 | 6 |
| 15 | 6 | 9 |
| 20 | 8 | 12 |
| 30 | 12 | 18 |
| 40 | 16 | 24 |
| 50 | 20 | 30 |
| 60 | 24 | 36 |

4. **If MPH class fails (Step 3):** Switch to an amphetamine. A reasonable sequence is to start with amphetamine sulfate (Evekeo), which is a milder amphetamine (equal parts d- and l- isomers)—about half the strength of Adderall per mg. For example, start Evekeo 5 mg once or twice daily and titrate up. If Evekeo or

mixed amphetamine salts (Adderall) are used, follow a similar weekly increase schedule as with MPH. Continue until the desired effect is achieved in two to three doses per day. (Note: There is currently no long-acting Evekeo, but long-acting liquid amphetamine options like Dyanavel XR or Adzenys ER can be used if pills are an issue.)

5. **If mixed amphetamines fail (Step 4):** Consider dextroamphetamine or lisdexamfetamine. In the rare case that a patient doesn't respond to any of the above stimulants, a trial of pure dextroamphetamine (e.g., Zenzedi IR 2.5–5 mg BID) can be tried. Another option is Vyvanse (lisdexamfetamine), a prodrug of dextroamphetamine. Vyvanse often is hit-or-miss in kids: some do extremely well, others do not respond as expected. Always start Vyvanse low (10 mg) and increase by 10 mg weekly; most children under 12 end up in the 20–30 mg range, rarely up to 50 mg. One advantage of Vyvanse is its flexibility in dosing: the powder from the capsule can be dissolved in water and then *part* of that solution given (e.g., half the solution to get an intermediate dose between 20 and 30 mg). This can help fine-tune doses like "15 mg" or "25 mg" which aren't otherwise available.

Throughout this process, if **two stimulant classes (MPH and AMP)** fail to produce adequate results, or side effects are limiting at every turn, it's reasonable to consider a **non-stimulant** (atomoxetine, guanfacine ER, etc.) or combination therapy. Additionally, always circle back to diagnosis and comorbidities—if the patient has a co-occurring condition (like anxiety, autism, tics, etc.), treating that or adjusting the ADHD plan in light of it may be necessary.

*(On the other hand, if a stimulant trial succeeds but with some side effects, you might consider augmentation with a non-stimulant rather than abandoning the stimulant. For example, adding low-dose guanfacine ER to a stimulant can help with aggression, impulsivity, or rebound phenomena. See below for augmentation strategies.)*

**Medication Formulation Tips** For very young or medication-sensitive children, some clinicians even start with extremely low doses like 2.5 mg of Ritalin in the morning and 5 mg at noon. While not every pharmacy can readily provide a 2.5 mg dose, cutting scored tablets or using a compounding pharmacy are options. In such cases, one might also use an intermediate product: Aptensio XR 10 mg (which releases ~4 mg immediately and 6 mg later) is an option if cost permits (an authorized generic exists). Metadate CD, another MPH XR with a 30/70 IR:ER ratio, could be used similarly (though it tends to last only ~6 h).

If tablets/capsules are a problem (child won't or can't swallow them), as mentioned, try alternative formulations or practice pill-swallowing techniques in the office. The availability of liquid (Quillivant XR), ODT (Cotempla), and patch (Daytrana) forms of stimulants provides substantial flexibility. Encourage families to also maintain routines like proper diet, as stimulants suppress midday appetite—ensuring a nutritious breakfast and dinner becomes important—and "medication holidays" on weekends or summers can be considered if growth is a concern (more on this in Sect. 5.2 Monitoring).

**Case Example (Wrap-Up)** After multiple trials, your patient is finally doing well on a combination of Adderall XR in the morning and a small Ritalin IR in late afternoon. The parents report marked improvement in homework completion and overall mood. However, they note that when he hit a growth spurt and entered middle school, the single morning capsule was no longer enough for the longer school day. This required a dose increase and eventually an afternoon booster. This underscores that ADHD medication plans are dynamic. Transition points—like moving from elementary to middle school (with later end times and more homework)—often necessitate re-evaluation of dose and schedule. It's not that the child became "tolerant" in the addictive sense; rather, the demands on them increased and their pharmacokinetics changed (bigger body, faster metabolism). It's rarely necessary to stop medication completely during school breaks (unless you're doing a planned trial off meds), but you might adjust the regimen (for instance, using just a short-acting tablet on days when the routine is light, to maintain some consistency and structure without over-medicating).

## 7.8 Key Differences in Prescribing for Children vs Adults

**Initiating Treatment in Children** Follow this simple stepwise approach in for initiating treatment for children:

**Start low, go slow** with methylphenidate. For instance, Ritalin 5 mg before school and 5 mg at lunch is a typical start. Then titrate upward in 5 mg increments. Once you find an effective IR dose schedule, consider switching to an equivalent long-acting MPH to improve convenience and compliance. Monitor not just ADHD symptoms but also appetite, sleep, and mood at each step. Only switch to amphetamine if MPH truly cannot achieve the desired effect or causes problematic side effects even at low doses.

**Adolescents (13–17 years)** often straddle the line between child and adult management. If an adolescent has been on stimulants since childhood, you may just continue their regimen and adjust as they grow (bearing in mind the puberty-related changes discussed). If an adolescent is newly diagnosed, many providers will start them on an amphetamine, as tolerability is usually better by this age and evidence (World Federation ADHD consensus) suggests amphetamines have edge in older teens. However, methylphenidate is still an appropriate choice for an adolescent, especially if there are any red flags for amphetamine (such as strong family history of addiction or patient anxiety about stimulant side effects). The algorithm for adolescents might be:

- If no prior treatment, consider starting with Adderall XR or Vyvanse for convenience (many teens prefer once-daily dosing). Starting doses could be Adderall XR 10 mg or Vyvanse 20 mg, with weekly increments.
- If the first stimulant is ineffective or not tolerated, try the other class (e.g., if Vyvanse didn't work, then try Concerta or Focalin XR).

- Pay extra attention to afternoon/evening coverage—many teens have homework, sports, jobs, etc. after school. They might need a booster or a longer-acting option like Mydayis (if early morning concentration isn't as critical, since Mydayis has a delayed onset).
- Involve the teen in decision-making. Adolescents are more likely to be non-adherent if they don't agree with the plan. Emphasize how medication can help with *their* goals (driving safely, getting into college, not feeling overwhelmed).
- Also, discuss stimulant diversion frankly. Many teens know classmates who misuse or share stimulants. Encourage them to never share their medication and to keep it secure (e.g., not leaving it in an unlocked locker or telling their friends or roommates where they keep their medications).

An important note for adolescents is to re-evaluate the diagnosis if things aren't adding up. Sometimes attention problems in teens could be due to emerging depression, substance use, or other issues. Ensure an ADHD diagnosis in adolescence meets full criteria and isn't better explained by something else, because stimulant treatment won't fix non-ADHD causes of inattention and could even worsen some conditions (like cause anxiety or mania in certain patients).

### 7.8.1 Important Treatment Risks with Teens and Adults

For adults, the approach is similar to adolescents with a few additional considerations: the preferred medication for treatment in adults is the amphetamine class. If an adult has no cardiac contraindications and no history of stimulant misuse, an amphetamine (such as Adderall XR or Vyvanse) is often first-line. If there is a concern such as the patient is very anxious, or has a relative who misused Adderall, one might start with methylphenidate instead.

**Dosing** Adults can often titrate a bit faster than children. It's not unusual to go up by 10 mg increments in an adult, whereas in a child we might do 5 mg increments. Just ensure the patient is tolerating each step.

**Schedule-II** Stimulant medications are classified as Schedule II controlled substances, requiring adherence to federal prescribing regulations governing storage, documentation, and prescribing practices (Drug Enforcement Administration 2023).

**Comorbidities** Many adults with ADHD have comorbid conditions (e.g., depression, anxiety, PTSD, substance use disorder). Treating ADHD can sometimes help those (e.g., reducing ADHD symptoms might lessen anxiety from constant failure), but be cautious: if a comorbidity is more severe, address it first. For example, an adult with uncontrolled bipolar disorder should not get stimulants until mood is stabilized, as stimulants could precipitate mania. On the other hand, prioritizing ADHD treatment first can be beneficial for some underling disorders such as Social

Anxiety. disorders. The persistent, self-referential nature of social anxiety—characterized by ongoing concern about how one is perceived by others—can significantly exacerbate executive dysfunction, working memory load, and emotional reactivity in individuals with ADHD, thereby influencing medication selection, particularly when choosing agents that reduce hyperarousal or improve emotional regulation ("Social anxiety is a persistent part of many people's thinking about how others would think about them") (Brown and Kennedy 2025)

**Misuse Risk** Always assess an adult's substance use history. If they have a history of stimulant abuse or are at high risk (e.g., active alcohol use disorder), consider non-stimulants like atomoxetine or lisdexamfetamine (Vyvanse) which has a lower abuse potential due to needing metabolic activation. Vyvanse and newer prodrugs (e.g., Azstarys which is serdexmethylphenidate/dexMPH) are designed to be harder to misuse (can't snort to get high easily). Document this rationale if you choose a prodrug for that reason. Stimulant treatment is associated with a small but measurable increased risk of psychosis, particularly with amphetamine formulations, underscoring the importance of careful monitoring in vulnerable populations (Moran et al. 2019).

**Safety/Driving** Advise adult patients to not perform safety-sensitive work (operating heavy machinery, driving long distances) until they know how the medication affects them. Generally, being on ADHD medication improves driving performance in those with ADHD, but initially one should be cautious until stable.

**Case Vignette (Adult)** Nicholas, whom we left off at age 12 having stopped meds, returns at 23 seeking treatment for ADHD as an adult. He had evaluations confirming persistent ADHD in adolescence and early adulthood. Now in college, he finds he needs help focusing. The NP and Nicholas decide to try Vyvanse 10 mg to start (due to its smoother profile and lower abuse potential). They instruct him to increase by 10 mg every few days as tolerated, up to 50–70 mg. Nicholas ends up finding 50 mg daily effective, but it only lasts ~6 h for him. Because of his long days on campus, the NP adds an afternoon dose of Adderall IR 10 mg, which Nicholas can take on days when he has late classes. This combination (long-acting in morning, short-acting later) allows him to fine-tune based on his schedule. Nicholas is counseled on not taking the afternoon dose if he doesn't need it (to avoid insomnia) and on safe storage of his meds (since he has roommates). He does well and reports improved academic performance.

Finally, starting stimulants should have a **pre-treatment checklist** of health items (see Appendix for a safety checklist). Key things: screen for cardiac symptoms (any chest pain, episodes of syncope, family history of sudden death), check blood pressure and pulse, and perhaps get baseline weight/BMI. If any cardiac risk factors exist, consider an EKG or cardiology consult before starting. Though serious cardiac events on stimulants are rare, it's prudent to be thorough. Also, document that you discussed with the adult the potential side effects (e.g., insomnia, appetite loss, sexual side effects like maybe a bit of ED or reduced libido on stimulants—some male patients report this, though it's not widely studied).

### 7.8.2 Process for Initiating and Fine-Tuning Treatment

For patients ~13 and older (adolescents and adults), clinical practice often shifts to favoring amphetamines as the first-line stimulant. By adolescence, many individuals can tolerate amphetamines well—and as discussed, biological changes during puberty can make amphetamines more effective than methylphenidate for some (particularly males). Therefore, an adolescent or adult who is newly diagnosed will frequently be started on an amphetamine medication (unless contraindications or specific reasons direct otherwise).

A typical starting point might be a generic XR of lisdexamfetamine or mixed amphetamine salts 10 mg once daily, taken 30 min before breakfast to minimize interference from appetite suppression and optimize absorption. Over 1–2 weeks, titrate each dose upward in 10 mg increments every few days until you achieve maximal symptom control with tolerable side effects. Many adolescents might stabilize at similar doses of adults. The exact dose is less important than the outcome—some may need as little as 5 mg, others 20 mg, depending on sensitivity.

Among short-acting stimulant options for teens/adults, there are essentially three choices: amphetamine sulfate (Evekeo), mixed amphetamine/dextroamphetamine (Adderall), and dextroamphetamine (Dexedrine). In terms of potency, Evekeo < Adderall < Dexedrine (because Evekeo has equal levo- and dextro-amphetamine, Adderall is 75% dextro/25% levo, and Dexedrine is 100% dextro which is the most potent isomer). In practice, Adderall IR is often the first choice for older patients due to its wide availability (generic on most insurance formularies), cost-effectiveness, and the convenience of multiple extended-release formulations (Adderall XR, Mydayis) that can be used later for fine-tuning.

If a patient finds Adderall helpful for focus but has side effects, they can't tolerate (e.g., moodiness or appetite loss), one strategy is to switch to the less potent formulation, amphetamine sulfate (Evekeo). For instance, if Adderall 5 or 10 mg was too much, you might try Evekeo 5 mg and later split the dose to two or three times a day to see if that provides a gentler effect. Conversely, if Adderall isn't effective enough at high doses, one could try dextroamphetamine sulfate (Zenzedi); although, typically if Adderall fails, then it's likely that a stronger formulation will not be more tolerable as dextroamphetamine IR is not drastically different.

In summary, systematically trialing different stimulants (perhaps two from the amphetamine class, plus methylphenidate class trials) is reasonable until either one is found to be helpful or all reasonable options are exhausted.

For **adult patients** specifically, medication selection also considers lifestyle and comorbidities. Amphetamines tend to be better tolerated in adults than in younger patients. Thus, starting with an XR amphetamine (Adderall XR or Vyvanse) is common as they are generic and covered on many insurance formulary. If an adult patient tries Vyvanse at the lowest dose (10 mg) and it isn't helpful and there are no side effects, it's usually better to increase the Vyvanse rather than switching to methylphenidate prematurely—as most adults will respond better to the amphetamine class if dosed adequately. After a fair trial of two different amphetamine formulations fail would we consider going to methylphenidate in an adult patient or consider the non-stimulants. (This is essentially the inverse of the pediatric algorithm.)

### 7.8.3 Non-stimulant Treatment in Children and Adults

Non-stimulant medications play a central role in the evidence-based treatment of ADHD across the lifespan. Although often labeled as "second-line," non-stimulants are essential alternatives—and in many cases, strategic adjuncts—to stimulant therapy. They are particularly valuable when stimulants are contraindicated, poorly tolerated, ineffective, or when specific comorbidities (e.g., anxiety, depression, autism spectrum disorder, tic disorders, or sleep problems) suggest an alternative mechanism of action may offer better functional benefit. Non-stimulants may also be preferred in situations of stimulant misuse risk, patient/family hesitancy, or when a smoother, less activating pharmacologic profile is desired. Furthermore, systematic review evidence suggests that pharmacologic treatment also improves associated behavioral outcomes, including oppositional symptoms and aggression, particularly with stimulant medications and alpha-2 agonists (Pringsheim et al. 2015).

The decision to initiate a non-stimulant should be individualized, rooted in shared decision-making, and based on a thorough review of ADHD symptoms, comorbid conditions, functional impairment, family/patient preference, and prior treatment response. Clinical practice guidelines from the American Academy of Pediatrics (AAP), American Academy of Child and Adolescent Psychiatry (AACAP), and expert-review literature converge on the principle that the goal of treatment is not merely symptom reduction but restoration of daily functioning and sustained improvement in quality of life. Non-stimulants frequently require longer titration periods and ongoing clinical follow-up to achieve optimal benefit. Taken together, the evidence supports a tiered pharmacologic approach in ADHD in which stimulants remain first-line due to superior efficacy, while non-stimulants serve as important alternatives or adjuncts based on tolerability, comorbidities, and patient-specific factors (Radonjić et al. 2023; Faraone et al. 2021).

### 7.8.4 Clinical Decision Pathway: Choosing a Non-stimulant

A practical visual framework is shown in Fig. 7.3 is a simplified flowchart which highlights two major pathways:

1. **Viloxazine ER (Qelbree)** for patients with ADHD complicated by depression, anxiety, autism, trauma history, stimulant sensitivity, or a contraindication to

stimulant treatment. Atomoxetine serves as a reasonable backup if viloxazine is ineffective or poorly tolerated.

2. **Guanfacine ER** for patients with combined-type ADHD, hyperactivity/impulsivity, sleep problems, irritability, stimulant rebound, substance use concerns, or co-occurring autism. Clonidine ER and Onyda-XR may be used as subsequent alternatives.

This approach aligns with guidelines emphasizing tailoring medication selection to neurocognitive profile, comorbidity pattern, and real-world functional needs. Non-stimulants are not merely substitutes for stimulants—they serve as targeted agents with distinct neurobiological mechanisms, unique benefits, and specific roles within multimodal ADHD care.

### 7.8.5 Atomoxetine (Selective Norepinephrine Reuptake Inhibitor)

*FDA-approved for adults and children ≥ 6 years*

Atomoxetine is a long-acting NRI with robust evidence for improving inattentive symptoms, emotional dysregulation, and executive function deficits. As highlighted in the uploaded document, atomoxetine increases both norepinephrine and dopamine levels in the prefrontal cortex without raising dopamine in the nucleus accumbens, limiting abuse potential.

**Indications** ADHD with comorbid anxiety, depression, or substance use risk; patients who cannot tolerate stimulants.

**Onset** Initial improvement in 2–4 weeks; full effect by 6–8 weeks.

- Dosing Adults: 40 mg/day → 80 mg/day → up to 100 mg/day
- Children: 0.5 → 1.2 mg/kg/day

- Clinical Pearls Slower titration with CYP2D6 inhibitors (fluoxetine/paroxetine) or known poor metabolizers.
- Monitor BP/HR, mood, and GI effects.
- Boxed warning: suicidal ideation., especially for patients with risk of bipolar disorder.
- Do not open capsules.
- May be taken with or without food, though absorption may be reduced.

clinicians should also be aware of cardiovascular precautions, growth monitoring, and the importance of early reevaluation within the first month of treatment.

### 7.8.6 Viloxazine ER (Qelbree)

*Norepinephrine modulating agent; FDA-approved for children and adults*

Viloxazine offers faster onset than atomoxetine (1–2 weeks) and has shown benefits in populations with anxiety, depression, irritability, and ASD traits.

**Dosing** 200 mg QAM → increase every 2-4 weeks to 400–600 mg/day. For children, adjust by 50% less by reducing starting dose to 100 mg and increase gradually in 2-4 week intervals to max of 300-400 mg/day.

**CYP Considerations** Potent CYP1A2 inhibitor; adjust meds such as tizanidine, ropinirole, and some antipsychotics.

**Monitoring** Appetite, insomnia, BP/HR.

### 7.8.7 Guanfacine ER (α-2A Adrenergic Agonist)

*FDA-approved ages 6–17; widely used off-label in adults*

Guanfacine ER reduces hyperactivity, impulsivity, emotional dysregulation, and stimulant rebound. It is well suited for ADHD with sleep problems, irritability, ASD, or tic disorders. As summarized in Table 7.4, guanfacine ER demonstrates a moderate effect size for ADHD symptom reduction. In the Cortese et al. network meta-analysis, guanfacine was superior to placebo in children and adolescents with a clinician-rated SMD of approximately −0.67, while pivotal placebo-controlled trials generally support effect sizes in the 0.4–0.6 range. Adult evidence is more limited, but one randomized, double-blind, placebo-controlled adult trial reported an effect size of 0.52, supporting guanfacine ER as a clinically meaningful non-stimulant option, particularly when stimulant tolerability, sleep, tics, emotional dysregulation, or adjunctive symptom control are relevant considerations.

**Dosing** 1 Mg Q AM or QHS → Increase every 4 weeks to max tolerated dose or 4 Mg)

**Pharmacology** CYP3A4 substrate; adjust dose with inhibitors/inducers.

**Precautions** Hypotension, sedation; taper slowly, by 1mg every 2-4 weeks to prevent rebound hypertension and orthostasis. Dry-mouth is a common and often overlapping side-effect when taken with stimulants.

### 7.8.8 Clonidine ER (α-2 Adrenergic Agonist)

Clonidine ER is useful for evening irritability, sleep onset problems, and hyperactivity/impulsivity.

**Dosing** Start 0.1 mg QHS → titrate by 0.1 mg every 5–7 days to 0.2–0.4 mg/day.

**Risks** Hypotension, bradycardia; always taper gradually by 0.1mg every 2-4 weeks to prevent rebound hypertension and orthostasis.

#### 7.8.8.1 Augmentation Strategies

Non-stimulants are effective adjuncts when stimulant monotherapy offers only partial benefit:

- Add guanfacine ER for persistent impulsivity, irritability, or stimulant rebound.
- Add clonidine ER for severe hyperactivity, insomnia, or aggressive outbursts.
- Use atomoxetine or viloxazine in patients with mood/anxiety comorbidities when stimulants alone are insufficient.

Pediatric RCTs support combination therapy with stimulants and α-2 agonists; adult evidence, while less extensive, is clinically supportive.

#### 7.8.8.2 When to Add an Alpha-2 Agonist

Adjunctive guanfacine ER or clonidine ER can:

- Reduce evening stimulant crash.
- Improve sleep initiation.
- Stabilize emotional dysregulation.
- Reduce hyperactivity/impulsivity not fully controlled by stimulants.

**Monitoring** BP, HR, sedation, driving safety.

#### 7.8.8.3 Summary

A structured, evidence-based approach to non-stimulant treatment allows clinicians to manage ADHD safely and effectively across complex clinical presentations. By incorporating expert guidelines, pharmacologic principles, titration pathways, and careful monitoring, nurse practitioners can deliver high-quality ADHD care that is tailored, safe, and grounded in functional improvement (Fig. 7.10).

### 7.8.9 Medication Transitions

#### 7.8.9.1 Clinical Tips During Puberty

The take-a-way from this section is to check and re-check dosage during growth spurts. If a pre-teenager is at the upper limit of MPH dosing (e.g., 40–60 mg per day) and starts to exhibit a diminished response to treatment, it's reasonable to stop the drug and consider a trial of an amphetamine rather than pushing the MPH dose higher. Also watch for new onset side effects like insomnia or mood lability in a previously stable patient—sometimes as metabolism speeds up, the dosing schedule might inadvertently be overshooting and causing addition side-effects, e.g., giving a second dose too late in the day causing insomnia, or dose too high causing evening irritability. Those signs can mean it's time to adjust timing or dose downward even as you consider a class switch.

Although stimulants remain among the most effective treatments for ADHD, clinicians often have concerns regarding their potential impact on growth and development. Findings from the ADHD Drugs Use Chronic Effects (ADDUCE) project suggest that methylphenidate treatment does not appear to substantially disrupt

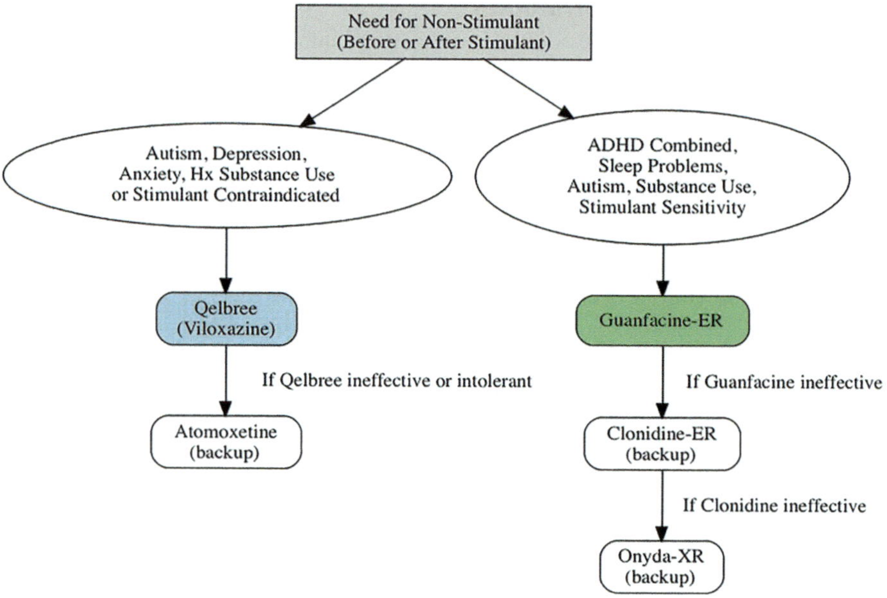


**Fig. 7.10** ADHD nonstimulant decision tree. (Created by Ryan J Kennedy)

pubertal maturation or bone age progression in most children and adolescents with ADHD (Carucci et al., 2024).

**Hormones** The rise in testosterone during puberty (and to a lesser extent, estrogen) speeds up dopamine clearance in the brain. More dopamine transporters become active under androgen influence, meaning a pure reuptake blocker like MPH might struggle to keep dopamine levels up. In clinical practice, parents, teachers, and patients may start to notice a gradual change in attention, regression to disruptive or inattentive behavior, and decrease in academic performance. This can be discussed with patients and their caregivers as the approach the years of puberty so the future adjustment in medication meets less resistance. As the dose of methylphenidate increases during this process, eventually the dose becomes less helpful, and a switch to the lowest dose of amphetamine can be initiated.

This *fades* during post-puberty effects methylphenidate's efficacy and should be expected from a biological perspective. With increased dopamine expression, changes in metabolism, and hormonal modulation of neurotransmitters, methylphenidate has limitations as it's primarily a dopamine reuptake inhibitor, and for this important reason for many adolescents, this drug progressively becomes less effective.

As adolescents grow, the effects of amphetamines become more tolerated and more effective as they have multiple mechanisms (blocking reuptake, reversing transporter flow, and promoting neurotransmitter release)—often become a better fit for teens and adults. Of course, cardiovascular safety and misuse risk must be considered with amphetamines, particularly in this age group, but those can be managed with proper screening and monitoring.

**Pharmacokinetic Changes** As adolescents grow taller and heavier, their volume of distribution increases, and their metabolic rate often accelerates. The same dose of medication now yields a lower blood concentration and shorter half-life than it did when they were smaller. It simple terms, preadolescent patients are "outgrowing" this class of stimulant.

These factors mean that around mid-adolescence, some patients need to switch from MPH to an amphetamine for optimal treatment. Clinically, one might notice a 14-year-old boy who needed 54 mg Concerta in 6th grade now needs something stronger in 8th grade. On the other hand, not all teens need a switch: often, girls who experience a later puberty and not a huge surge of testosterone can maintain well on methylphenidate longer, whereas boys with an early big testosterone increase are the "early switchers" to amphetamine.

Therefore, switching to an amphetamine reignites dopamine transporters by additional pathways which release dopamine and additionally block its reuptake, thus bypassing the up regulation of dopamine transporters to a clinically meaningful degree.

### 7.8.10 Girls and Women

Estrogen plays a crucial role in regulating dopamine synthesis within the prefrontal cortex. For girls and women with ADHD, fluctuations in estrogen levels can exacerbate the manifestation of ADHD symptoms, particularly considering the natural fluctuations in estrogen levels throughout the menstrual cycle. Sex hormones appear to play an important role in the regulation of attention and cognitive functioning. Research examining the effects of 17β-estradiol has demonstrated measurable impacts on attentional performance, highlighting the potential influence of hormonal fluctuations on ADHD symptom expression in females (Chen et al., 2016). This aspect warrants discussion during medication optimization, where it is essential to engage with girls and women about their ability to manage their cognitive and emotional functioning within the current treatment regimen. An experimental study conducted by experts in the Netherlands noted female patients in their clinic with "diminished response to amphetamines in the late luteal phase" and noted significant benefits in ADHD symptoms and mood regulation by increasing premenstrual psychostimulant dosage (de Jong 2023).

Consider this case example where your 20-year-old female college student who takes 30 mg of a long-acting amphetamine in the morning before college courses scheduled on weekdays. However, a noticeable pattern occurs each month where she struggles with a transient decrease with normally effective ADHD medication that works perfectly well until the premenstrual luteal phase of her monthly cycle. Since she is taking a capsule 5 days per week (25 capsules per month) and not taking any on the weekends, it would be appropriate to provide her with an additional prescription, for 40 mg to boost dopamine signals caused by a decrease in estrogen levels ensuring she still receives 30 capsules for 30 days. This adjustment would allow for a slight increase in the dosage to cover the week when ADHD symptoms are more pronounced.

### 7.8.11 Monitoring Response to Treatment

Once medication is underway, **close monitoring** is essential, especially in the first few weeks of titration. The goal in early follow-up is to find that "sweet spot" dose that gives maximal therapeutic benefit with minimal side effects. Expect that side effects are **common during initial titration**—most are dose dependent and often transient (they may subside after the first 3–5 days at a new dose as the body adjusts). Common stimulant side effects include reduced appetite, insomnia, anxiety or nervousness, stomachache, headache, mild increase in heart rate or blood pressure, dry mouth, and (less commonly) motor tics. Families often worry about growth suppression; studies suggest any effect on height is *very small* (on the order of a centimeter or two of difference in final adult height, and even that is debated). If growth is a concern, one strategy is to have medication holidays on weekends or summers, though this must be balanced against whether the patient needs daily symptom control for safety (e.g., an adolescent driving a car might need medication daily, not just on school days). You might reassure parents that many kids take stimulant breaks over summer with no issue, but others actually benefit from staying on medication year round (for structure, self-esteem, safer driving, etc.).

One practical tool for follow-up is a **structured questionnaire or template** to review symptoms and side effects. Below is an example of an acronym to use for a patient ADHD medication appointment template that can be used during follow-up visits (or via phone/email):

**F.I.N.E.-T.U.N.E.**
F—Focus improvement
I—Issues or side effects
N—Noticed timing of onset
E—Effect duration
T—Timing of administration
U—Unwanted symptoms at wear-off
N—Needed adjustments
E—Expected goals

### 7.8.12 Addressing Side Effects

Despite our best efforts at careful titration, side effects can still occur and sometimes cause patients or parents to question the medication. The key is to anticipate and manage side effects proactively so they don't derail treatment.

#### 7.8.12.1 "Stimulant Crash" or Rebound

One phenomenon to educate families about is the potential rebound effect when a stimulant dose wears off. A "stimulant crash" refers to the patient becoming withdrawn, tired, irritable, or moody as the medication level drops. This often happens in the mid to late afternoon for long-acting capsules—coincidentally when most kids are coming home from school and when metabolism naturally peaks, potentially exacerbating the effect. Younger children and those with underlying anxiety or mood sensitivities are more prone to noticeable rebound.

It's important to normalize this and plan for it: "Around 4 or 5 PM, you might see him get weepy or touchy as the medicine wears off. This doesn't mean he's 'addicted' or anything; it's a known effect of it leaving his system quickly." Explain that rebound is common and manageable, so patients don't quit their medication prematurely out of alarm.

**Several Strategies Can Alleviate Rebound** Assess for problems with irregularities with sleep, appetite, hydration, stress, or other medical or social problems. For example, not feeling hungry because of a stimulant or *hyper-focus* on tasks and skipping meals worsen the late-day crash.

Consider adjusting the dosage or formulation: sometimes a lower morning dose results in a gentler drop-off, or using brand-name products or a longer-acting formulation instead of immediate-release tablets in the afternoon yields a smoother decline.

A very effective strategy is a small short-acting dose in the afternoon, given about 30 min before the expected crash. For example, if a child on Concerta or Vyvanse typically crashes at 4 PM, taking a 5 mg Ritalin or Adderall at 3:30 PM can bridge the gap and let the stimulant wear off more gradually. This often prevents the sharp contrast that triggers rebound behaviors. The analogy I often use that patients readily understand is that we want the medication to wear off smooth like a plane preparing for landing compared to the quick decline of a helicopter.

In some cases, adding a non-stimulant (like an alpha-2 agonist) can smooth out evenings. For instance, guanfacine ER or Qelbree given in the morning or early afternoon can help with irritability and hyperactivity into the evening and may improve the transition off stimulant.

Always discuss these possibilities *before* starting the medication and periodically during follow up visits, so families know to report rebound and not just stop the drug. Reiterate that rebound symptoms are not the patient's lack of willpower in regulating their behavior; instead, it's a transient medication effect that can be managed.

If a patient experiences excessive stimulation side effects (e.g., feeling "too jittery" or anxious shortly after a dose), that indicates the dose might be too high. Remind them that feeling like they had "too many cups of coffee" is a sign to dial the dose down a bit. Fine-tuning in those first weeks is normal.

**Adjustments Over Time** A well-designed stimulant regimen generally does not cause *physiological tolerance* (i.e., needing ever-higher doses due to diminished effect). However, life changes and developmental stages often require dose adjustments. As mentioned, a child moving to middle school or a teen starting a job may need a higher dose or an extra dose for new challenges. Conversely, during summer break or periods of low stress, the dose might be dialed down. It's not always necessary (or desirable) to stop medication during these breaks—keeping some medication on board can help maintain the habits and routines that were built during treatment. For example, if a teen has done well organizing themselves on meds, a summer off meds could see some regression; instead, using just a small morning dose in summer might keep them in a good routine without overmedicating when demands are lower.

If stimulants are not helping despite adequate trials, or if the patient has a comorbidity that complicates stimulant use (like significant anxiety, tic disorder, or a history of substance misuse), then involving a specialist or colleague (such as a child psychiatrist or a psychopharmacology consultant) can be very helpful. Sometimes a fresh perspective can identify a missed issue (e.g., the diagnosis might need revisiting, or perhaps a combination therapy is warranted). Collaboration ensures the patient gets the benefit of broader experience and helps the primary provider feel supported in tough cases.

#### 7.8.12.2 Side Effect Management Recap

Common side effects and their management can be summarized to patients as follows:

**Decreased Appetite** This usually affects lunch more than dinner. We'll monitor growth. Focus on an adequate and healthy balanced breakfast and dinner. If needed, you can reduce the dose or switch to a shorter-acting form to help with appetite by controlling the duration with multiple tablets per day to avoid interference with meals. For example, taking a 10 mg tablet just before breakfast and lunchtime.

**Sleep Changes** If this happens, let me know. You might need to give the dose earlier, lower the afternoon dose, or in some cases use a low-dose melatonin or change medication. Ask questions to ensure good sleep hygiene too.

**Mood Changes** This can happen when the medicine is wearing off (the "crash") or if the dose is a bit high. There are sometimes simple ways to handle this problem, often by adding a "booster dose" as a small immediate-release dose in the afternoon, or start by adjusting the morning primary dosage or delivery system.

**Headaches or Stomachache** These side effects are the most common in children and often get better after the first few weeks of adjustment to a new medication. Make sure their medication is taken with food, and remind parents that this also will delay the onset of action for most stimulants by an hour or 2.

**Hydration** If headaches persist, first ask about the amount of water and hydration before trying a different medication.

**Tics** Motor tics are rare as a side-effect, typically in children; however, this may likely be the dosage is too strong, or the class of medication needs to be switched. Stimulants can unmask a tic in a small number of kids. If that happens to persist, a helpful strategy is to stop the stimulant for a week or 2 while starting a non-stimulant to help reduce the sensitivity to the stimulant. Often this mitigates or stops the tic completely. If the tics persist concurrent to modifications to dose, then consider the tic to be a secondary condition that requires further assessment Most of the time, stimulants don't cause new tics in adults, but watch for signs of increased signs of stress that may manifest.

## Appendices

### Follow-Up ADHD Stimulant Treatment Assessment Template

#### Onset of Action

- Is the medication initiating effects at an appropriate time?
- Is effective focus achieved within approximately 30–60 min of the dose?
- Is the onset well tolerated?
  - Any surge of anxiety or "wired" sensation?
- Are there side effects soon after administration?
  - At what time of day do early side effects occur, if any?
  - Do these side effects resolve as the day progresses or persist?
- Is the medication taken consistently as prescribed (same time each day)?

#### Side Effects Related to Dosage

- Are there early side effects suggesting the dose may be excessive?
- Are there side effects suggesting a potentially inappropriate medication class (e.g., moodiness with amphetamine versus methylphenidate)?

#### Duration of Action

- What is the approximate duration of each dose's effect?
- What is the average duration of benefit?
- Is there a distinct period during which the medication effect has clearly diminished ("trough")?
  - If present, are ADHD symptoms returning during this period?
- Are there side effects that occur only when the dose wears off (rebound effects)?
  - If yes, describe these rebound effects.

#### Efficacy

- How would the overall symptom control be rated at this dose?
- In which areas or time periods is ADHD still interfering with function?
- Are there any instances where the patient is "too quiet" or excessively focused, suggesting potential over-medication?

#### Sleep

- Has there been a change in sleep quality since starting this medication?
- Is there difficulty initiating sleep?
- Any changes in bedtime or increased tendency to stay awake later?

- Are there nighttime awakenings?
- Are there early morning awakenings (waking significantly earlier than desired)?
- What is the typical wake time in the morning?
- Is it difficult to get out of bed (morning grogginess)?

### Mood

- How is the overall mood on medication days?
- Are there episodes of irritability, tearfulness, or increased emotional sensitivity?
  - If so, when do these occur?
- Does mood variability correspond with dosing times?

### Appetite

- What is the overall appetite level?
- Which meals are most affected?

### Stress and Anxiety

- Have stress levels or anxiety symptoms changed since starting or adjusting the medication?

### Physical Activity

- What is the energy level for exercise or physical activity?
- Is the patient maintaining an active lifestyle?

### General Health

- Any new changes in general health status since the last visit (e.g., new illnesses, new medications)?
- Current weight.
- Any new or worsening headaches, blood pressure concerns, or other physical complain.

## Practitioner's Controlled Substance Checklist

| Checklist item | Details for safe prescribing practices |
|---|---|
| Assess baseline | Screen for contraindications such as |
| Assess comorbidity | Address mood, anxiety, or substance use issues either first or in parallel |
| Review recommendations | Ensure first-line meds considered or tried before off-label, especially for children and special populations |
| Informed consent | Ensure patient/parents understand rationale and risks/benefits of off-label or higher dose |

| Checklist item | Details for safe prescribing practices |
|---|---|
| Monitoring plan<br>Office and telehealth policies | Set up follow-up to check response and vitals, monitor blood pressure, pulse, weight, and height<br>Use of a patient portal for communication is also strongly recommended, which can make the use of follow-up rating scales simpler, enhance adherence, smooth prescription refill request |
| Documentation | Write down rationale for any deviation from standard practice |
| Safe and legal e-prescribing | Always first check before prescribing CS medication and documenting that you reviewed your state's "prescription drug monitoring database" (PDMD). If a patient was seen by another provider in another state request this information from the previous provider or obtaining permission to create a profile for that state<br>Each state may have individual requirements for scheduled follow-up for Schedule II medications used to treat ADHD<br>HEDIS (healthcare effectiveness data and information set) requirements usually mandate a follow-up appointment within 30-days after starting a new medication and follow-up should be scheduled within the next 30–60 days depending on the situation<br>Additional visits should be required for significant changes; however, it is possible to provide instructions for titration as long as it is clearly documented and a follow-up note is provided for continuity in the plan of care<br>Once the provider and patient agree that a treatment is stable, patients should be scheduled at least once every 90 days, or before they run out of their medication, if given up to a max of a single 90-day prescription to mail-order pharmacy and/or before renewing a series of 3 individual Rx's that cover 3 months |

Reference
deadiversion.usdoj.gov/GDP/(DEA-DC-071)(EO-DEA226)_Practitioner's_Manual_(final).pdf#page=53.10

## Pre-stimulant Treatment Safety Checklist

Before initiating a stimulant, review the following with the patient (and/or parent) and document that these items were assessed:

Psychiatric History History of a mood disorder or bipolar disorder? ☐ Yes / ☐ No

- History of psychosis or schizophrenia? ☐ Yes / ☐ No
- History of Tourette's syndrome or motor/vocal tics? ☐ Yes / ☐ No
- History of substance use disorder (drugs or alcohol)? ☐ Yes / ☐ No
- Known allergy or serious adverse reaction to stimulant medications? ☐ Yes / ☐ No*(Rationale: Stimulants can exacerbate bipolar or psychotic symptoms, can worsen tics in some cases, and have abuse potential in those with substance use disorders. If "Yes" to any of these, proceed with caution or consider non-stimulants/consultation.)*

- Cardiac and Physical Health History of fainting or syncope episodes? □ Yes / □ No
- History of unexplained dizziness (particularly with exertion)? □ Yes / □ No
- History of chest pain, arrhythmia, or heart racing? □ Yes / □ No
- History of shortness of breath (unrelated to asthma) on exertion? □ Yes / □ No
- Family history of **sudden cardiac death** or serious arrhythmia (before age 50)? □ Yes / □ No
- Personal history of any diagnosed heart condition (murmur, cardiomyopathy, etc.)? □ Yes / □ No
- Ever had an abnormal EKG or been told of a heart rhythm issue? □ Yes / □ No
- History of exercise intolerance (unable to keep up with peers due to fatigue)? □ Yes / □ No*(Rationale: These screen for potential underlying cardiac issues that stimulants could aggravate. If any "Yes," an evaluation by a cardiologist or an EKG might be warranted before starting stimulant.)*

If all answers are **No**, the risk of a serious cardiac event on stimulants is extremely low, and one can proceed with standard care (while still monitoring vitals during treatment). If any answers are **Yes**, further investigation is needed (e.g., obtain an EKG, get cardiology clearance) prior to prescribing a stimulant.

## Sample Prior Authorization Documentation

Insurance companies often require **prior authorization (PA)** for certain ADHD medications, especially brand-name or high-dose therapies. They usually ask for documentation of trials of preferred alternatives and reasons for switching. Below are sample templates illustrating information to include for a PA request:

### A. Stimulant Prior Authorization: Requesting Dyanavel XR

- Document at least three past medication trials of equivalent XR drugs within the same class, with doses, outcome, and summarize with a statement for medical necessity or rationale, with any supporting evidence.

1. **Adderall XR**—Dosages tried: 10 mg, 20 mg, 30 mg IR twice daily; *Duration:* 6 months (Jan–June 2023).

**Outcome:** 10 mg was ineffective; 20 mg provided partial day coverage but required supplemental IR dosing in afternoon; 30 mg IR caused recurrent irritability due to the biphasic peak (early wear-off leading to mood lability). (This indicates we need an XR formulation for smoother release.)

2. **Vyvanse (lisdexamfetamine)**—Dosages tried: 10 mg, 20 mg, 40 mg; *Duration:* 12 weeks (July–Sept 2024).

**Outcome:** Inconsistent efficacy and inadequate duration. (Patient would focus in morning but lose effect by early afternoon, even at max tolerated dose.)

3. **Mydayis (mixed amphetamine salts triple-bead)**—Dosages tried: 12.5 mg, 37.5 mg; *Duration:* 3 months (2023).

**Outcome:** Excessive duration interfered with appetite and sleep (even lowest dose lasted too long for patient's needs).

- **Rationale for Requested Med:** Adderall XR provides a midpoint option: longer coverage than IR but shorter than Mydayis. Given the above trials, Adderall XR 25 mg is expected to cover the work day without late-evening effect, and the patient tolerated Adderall IR well aside from the short duration. We seek coverage for Adderall XR as the next logical step.

### B. Non-stimulant Prior Authorization: Requesting Qelbree

- Document at least two past medication trials of equivalent drug within the same class, with doses, outcome, and summarize with a statement for medical necessity or rationale, with any supporting evidence.

1. **Guanfacine ER (Intuniv)**—Dosages: 1–3 mg daily; *Duration:* 6 months (2023).
   - **Outcome:** Insufficient symptom improvement at 1 and 2 mg daily; maximum tolerated dose (3 mg) caused blood pressure to drop significantly at 3 mg trial, which was subtherapeutic.
2. **Atomoxetine (Strattera)**—Dosages: 20 and 40 mg daily; *Duration:* 6 months (2021).
   - **Outcome:** Insufficient symptom improvement at maximum tolerated dose (40 mg). Patient's blood pressure dropped significantly at 80 mg attempt, so 40 mg was highest dose tolerated, which was subtherapeutic.

- **Rationale:** Patient has hypertension and cannot tolerate high-dose stimulant. Guanfacine ER is being used off-label for adult ADHD in this case due to its complementary mechanism and mild blood pressure lowering (which is beneficial for this patient). Two small RCTs support guanfacine ER efficacy in adults. The patient has tried atomoxetine without success. We request coverage for guanfacine ER 2 mg daily.

*(If required, attach supporting literature:* e.g., *Iwanami* et al. (2020*) study showing guanfacine ER efficacy in adults.)*

### C. BRAND ONLY, Evidence-Based Justification: Xelstrym XR

Sometimes a PA is required even if the patient hasn't tried two or more other generics first, especially for unique situations. In such cases, provide a scholarly justification. For example: "This patient has a documented history of substance use. This is a request for Xelstrym XR (transdermal patch) because this drug's delivery system is a type of pro-drug stimulant and has lower abuse potential—requiring transdermal versus metabolic activation, it cannot be readily misused intravenously or intranasally. Studies (e.g., *Journal of Addiction Medicine, 2017; CNS Drugs, 2016*) demonstrate reduced liking and abuse-related events with compared to immediate-release amphetamines. Given the patient's history, it is in the interest of safety to use Xelstrym or Vyvanse as first-line rather than a short-acting amphetamine or mixed-amphetamine salts—extended release."

*(If available, cite specific studies or guidelines: "e.g., article/year shows lower diversion rates with lisdexamfetamine.")*

Always include: Medication name, dose, duration of each trial, and why it was stopped (lack of efficacy or intolerable side effect). For non-stimulants, note if they were used adjunctively or solo. If you haven't done two trials (e.g., patient has a contraindication), explain that "Patient has severe tic disorder, so we avoided amphetamine trials and request Strattera directly, per Tourette's guideline".

## Dose Conversion Reference

Because patients may switch between different stimulant formulations, it's helpful to know approximate dose equivalencies. Below are reference points for converting between some common medications (these are *approximations*, individual response varies):

**Amphetamine Equivalents:** (Using Adderall XR as baseline)

- Adderall XR 20 mg **Adzenys XR-ODT 12.5 mg** (Adzenys ODT is ~63% the potency of Adderall XR).
- Adderall XR 20 mg **Evekeo 26 mg** (Amphetamine sulfate requires ~1.3× the dose).
- Adderall XR 20 mg **Dexedrine Spansule 16 mg** (Dextroamphetamine ER is ~0.8×; dextro is more potent per mg).
- Adderall XR 20 mg **Vyvanse ~ 50 mg** (Lisdexamfetamine 50 mg converts to ~15 mg dextroamphetamine base = roughly Adderall XR 20 mg; Vyvanse 70 mg ~ Adderall XR 30 mg).*(Conversion factors: Adzenys XR-ODT is 0.63×, Evekeo 1.3×, Dexedrine XR 0.8×, Vyvanse ~ 2.6× by mg, relative to Adderall XR).*

**Methylphenidate Equivalents:** (Using Ritalin LA 40 mg as baseline)

- Ritalin LA 40 mg **Concerta 48 mg** (Concerta OROS has slightly higher mg for equivalent effect—about 1.2×).
- Ritalin LA 40 mg **Cotempla XR-ODT ~ 34 mg** (Cotempla ODT is ~0.85×; 34.4 mg ODT equals ~40 mg LA).
- Ritalin LA 40 mg **Jornay PM 52 mg** (Jornay needs ~1.3× dose to cover full day with delay).
- Ritalin LA 40 mg **Focalin XR 20 mg** (Dexmethylphenidate is about twice as potent, so half the mg).*(Conversion factors: Concerta ×1.2, Cotempla ×0.86, Jornay ×1.35, Focalin XR (dexMPH) ×0.5 relative to Ritalin LA).*

For more extensive conversion needs, refer to tools like the "Switching Stimulants Dose Conversion Calculator" (available at psychopharmacopeia.com).

Remember these are ballpark guides; always titrate based on clinical response. When switching, start a bit lower than the "equivalent" and adjust as needed.

## Resources for Clinicians

- **The Carlat Report (newsletter and podcast).**
- **The ADHD Report:** bimonthly newsletter edited by Dr. Russell Barkley, featuring research updates and clinical insights on ADHD. (Guilford Publications; subscribe via 1–800–365–7006 or at Guilford's website.)
- **ADHD Medication Guide:** A visual laminated chart of all FDA-approved ADHD medications, useful for patient education on different options. (Available at adhdmedicationguide.com).
- **FDA MedWatch Reporting:** Encourage reporting of any serious or unexpected medication side effects. Clinicians and patients can use the FDA's MedWatch Online Voluntary Reporting Form (find it on the FDA website) to report adverse events. *Note:* Adverse events can range from common issues like skin rashes to rare serious complications—reporting helps the FDA monitor safety in the population.
- **Patient Organizational Tools:** Recommend families keep a daily log or journal of the patient's routine (sleep, medication times, behavior, mood, diet). Tools like daily report cards or apps can help track patterns and make follow-up discussions more data-driven.
- **Collaboration and Follow-up:** Connect with local CHADD (Children and Adults with ADHD) chapters or online NP forums for ADHD to share experiences. For complicated cases, don't hesitate to refer to or co-manage with a psychiatrist or developmental-behavioral pediatrician.

By utilizing these resources and maintaining a learner's mindset, clinicians can stay at the forefront of ADHD treatment and provide the best care to their patients.

## References

Aiken C, Feder A, Carlat D. Prescribing psychotropics: from drug interactions to genetics. Newburyport: Carlat Publishing; 2022.

American Academy of Pediatrics. Clinical practice guideline for the diagnosis, evaluation, and treatment of ADHD in children and adolescents. Pediatrics. 2019;144(4):e20192528.

American Psychiatric Association. Diagnostic and statistical manual of mental disorders. 5th ed. Washington, DC: APA; 2013.

Arnsten AF. The emerging neurobiology of attention deficit hyperactivity disorder: the key role of the prefrontal association cortex. J Pediatr. 2009;154(5):I-S43. https://doi.org/10.1016/j.jpeds.2009.01.018.

Barkley RA. Attention-deficit hyperactivity disorder: a handbook for diagnosis and treatment. 4th ed. New York: Guilford Press; 2014.

Bishop JR, Zhou C, Gaedigk A, et al. Dopamine transporter and CYP2D6 gene relationships with ADHD treatment response in the methylphenidate vs. Atomoxetine crossover study. J Child Adolesc Psychopharmacol. 2024;34(10):458–69.

Brown TE. A new understanding of ADHD in children and adults: Executive function impairments. Routledge. 2013.

Brown TE, Kennedy RJ. Social anxiety: Hidden fears and shame in teens and adults. New York, NY: Routledge. 2025.

Brown TE, Kennedy RJ. (2021). Establishing a psychosocial plan to manage ADHD. In S. V. Faraone (Ed.), The World Federation of ADHD guide. 2021:63. Available from https://www.russellbarkley.org/factsheets/TheWorldFederationOfAdhdGuide.pdf

Carucci S, Zuddas A, Lampis A, et al. The impact of methylphenidate on pubertal maturation and bone age in ADHD children and adolescents: results from the ADHD drugs use chronic effects (ADDUCE) project. J Atten Disord. 2024;28(5):722–39.

Chen S, Kolangara S, Mokler D. How does estrogen influence attention? Impact of 17-beta estradiol on performance in an attentional set shifting task. FASEB J. 2016;30 https://doi.org/10.1096/fasebj.30.1_supplement.1284.2. (Abstract).

Connor DF, Findling RL, Kollins SH, et al. Effects of guanfacine extended release on oppositional symptoms in children aged 6–12 years with ADHD and oppositional symptoms. CNS Drugs. 2010;24(9):755–68.

Cortese S, Adamo N, Del Giovane C, et al. Comparative efficacy and tolerability of medications for ADHD in children, adolescents, and adults: a systematic review and network meta-analysis. Lancet Psychiatry. 2018;5(9):727–38.

de Jong M, Wynchank D, van Andel E, Beekman F, Kooij S. Female-specific pharmacotherapy in ADHD: premenstrual adjustment of psychostimulant dosage. Front Psych. 2023;14:1306194.

Drug Enforcement Administration (2023) Practitioner's manual: an informational outline of the controlled substances act. DEA-DC-071 (EO-DEA226). [Online]. Available at: https://www.deadiversion.usdoj.gov/GDP/(DEA-DC-071)(EO-DEA226)Practitioner's_Manual(final).pdf

Faraone SV. The pharmacology of amphetamine and methylphenidate: relevance to the neurobiology of ADHD and other psychiatric comorbidities. Neurosci Biobehav Rev. 2018;87:255–70.

Faraone SV, Banaschewski T, Coghill D, et al. The World Federation of ADHD International Consensus Statement: 208 evidencebased conclusions about the disorder. Neurosci Biobehav Rev. 2021;128:789–818. https://doi.org/10.1016/j.neubiorev.2021.01.022

Farhat LC, Flores JM, Avila-Quintero VJ, et al. Treatment outcomes with licensed and unlicensed stimulant doses for adults with ADHD: a systematic review and meta-analysis. JAMA Psychiatry. 2024;81(2):157–66.

Iwanami A, Saito K, Fujiwara M, et al. Efficacy and safety of guanfacine ER in the treatment of ADHD in adults: results of a randomized, double-blind, placebo-controlled study. J Clin Psychiatry. 2020;81(3):e1–8.

Lenzi F, Cortese S, Harris J, Masi G. Pharmacotherapy of emotional dysregulation in adults with ADHD: a systematic review and meta-analysis. Neurosci Biobehav Rev. 2018;84:359–67.

Moran LV, et al. Psychosis with methylphenidate vs amphetamine treatment of ADHD. N Engl J Med. 2019;380:1128–38.

Moukhtarian TR, Cooper RE, Vassos E, et al. Effects of stimulants and atomoxetine on emotional lability in adults with ADHD: a systematic review and meta-analysis. Eur Psychiatry. 2017;44:198–207.

Pringsheim T, Hirsch L, Gardner D, Gorman DA. Pharmacological management of oppositional behavior, conduct problems, and aggression in children and adolescents with ADHD, ODD, and CD: a systematic review and meta-analysis. Part 1: psychostimulants, alpha-2 agonists, and atomoxetine. Can J Psychiatr. 2015;60(2):42–51.

Radonjić NV, Bellato A, Khoury NM, Cortese S, Faraone SV. Nonstimulant medications for ADHD in adults: systematic review and meta-analysis. CNS Drugs. 2023;37(5):381–97.

Society for Developmental and Behavioral Pediatrics. Clinical practice guideline for the assessment and treatment of attention-deficit/hyperactivity disorder in children and adolescents [Internet]. Itasca (IL): Society for Developmental and Behavioral Pediatrics; 2019 [cited 2026 May 3]. Available from: https://www.sdbp.org/adhd-guideline.

Stahl SM. Mechanism of action of stimulants in ADHD: dopamine and norepinephrine reuptake inhibition. J Clin Psychiatry. 2010;71(1):12–3.

U.S. Food and Drug Administration. Adderall XR (mixed amphetamine salts) prescribing information. Current label; 2025a

U.S. Food and Drug Administration. Authorized generics. FDA consumer information page. 2026. Access Date: May 3, 2026. https://www.fda.gov/drugs/abbreviated-new-drug-application-anda/fda-list-authorized-generic-drugs
U.S. Food and Drug Administration. Adzenys XR-ODT (amphetamine) prescribing information. Current label; 2025b
U.S. Food and Drug Administration. Vyvanse (lisdexamfetamine dimesylate) prescribing information. Current label; 2025c

# 8 Navigating Comorbidities and Complex Cases of ADHD

Helene Vossos

## 8.1 Introduction

ADHD co-occurring disorders have a prevalence of 60% to 100% in children with one or more disorders continuing into adulthood (Gnanavel et al. 2019). The majority of the research conducted in the United States has been with children and adolescents. This includes 42% of children with autism spectrum disorder (ASD) and ADHD and includes 17% with a learning disability (LD) which does continue into adulthood (Gnanavel et al. 2019). It has been studied in children and adolescents with ASD and ADHD concurrently, symptoms overlap with more severe symptomatology and pervasive severe sleep-related symptoms (Gnanavel et al. 2019). More specifically, incidence with ADHD is 59% have ASD, 70% have LD, 50% have ODD, 50% have depressive disorders, 35% have at least one anxiety disorder, 55% have tic disorders, 10% conduct disorders, and 57% bipolar mood disorders (Gnanavel et al. 2019). These psychiatric comorbidities may impede adulthood growth and development if not diagnosed and treated. The diagnostic challenge for clinicians is the overlapping symptoms. Historically, the later the diagnosis, the poorer the outcome.

## 8.2 Definition and Etiology of Comorbidities of ADHD

Adults with attention-deficit disorder (ADD) comorbidities include 52% who have at least one other psychiatric disorder and who 40% have multiple comorbidities (Ohnishi et al. 2019). ADHD data has been researched revealing approximately 75% of ADHD cases have inheritability linkage (Borland et al. 2022). Many of the ADHD cases were born from a mother prematurely or subsequently had an

H. Vossos (✉)
University of St. Augustine for Health Sciences, St. Augustine, FL, USA

© The Author(s), under exclusive license to Springer Nature Switzerland AG 2026
R. J. Kennedy (ed.), *Assessment & Treatment of ADHD*,
https://doi.org/10.1007/978-3-032-24171-9_8

infection (Borland et al. 2022, p. 139). This data includes similar overlapping etiology of adults with learning disorders, autism spectrum disorders, or tic disorders. Clustering of symptoms and patterns include similar clinical manifestations, although working up a differential diagnosis list is essential.

Individuals with ADHD and conduct disorder or bipolar affective disorder in adults also have impaired functioning with mathematics (Borland et al. 2022). Overlapping symptoms of learning impairment, dysfunctional social functioning, and affective or impulse symptoms could affect academic or work performance.

## 8.3 Diagnosis of Common Comorbidities of ADHD and Case Studies

### 8.3.1 Autism Spectrum Disorder and Other Learning Disorders

Autism spectrum disorder (ASD) in 1943 historically was described in 11 case studies by Dr. Leo Kanner which highlighted the symptoms of "inability to relate" to other people in usual ways (Sadock et al. 2009, p. 3540). Early identification of ASD in the United States initially held a prevalence rate of 4.5 (450 cases) in 10,000 (1966); by 2020 prevalence rate was identified by 1 in 36 children (278 cases, in 2020) (CDC, Autism and Developmental Disabilities Monitoring Network (ADDM) 2025). The lower prevalence rates are postulated to be lower secondary to larger population pool, other pervasive development disorders, and other categories of autism spectrum expressive disorders.

Learning disorders encompass the etiology of genetic linkage and environmental factors. In diagnosing learning disorders, clinicians should assess for symptoms that start in childhood and are persistent, and findings include pervasive difficulties in learning academic skills. Academic skills such as language, mathematics, expression in spelling, fluency in writing, and reading may be affected in more than one domain compared to children or adolescents in similar academic settings. The criteria for a specific learning disorder diagnosis is the time of 6 months or longer of symptoms, academic reading, spelling, mathematics, or mixed scholastic skill impairment (Borland et al. 2022). In 1975, The Department of Disabilities Education Act was implemented in the United States charged with integrating appropriate and free resources for special education academic services (Borland et al. 2022).

## 8.4 Tic (Tourette's) and Motor Disorders

Tics or Tourette's disorders are considered a neuropsychiatric disorder characterized by rapid, brief episodes of motor or vocal movements (Borland et al. 2022). Tic (Tourette's) disorder includes motor or vocal tics that that last over 12 months, including one or more motor of vocal tic, and exclusion criteria would be if an individual was impaired from medication, substances, or another medical illness

(DSM-5, TR 2022). Tourette's disorder is characterized by the number and frequency of tics and vocalizations and initially may develop on the face or neck, most commonly effecting head, face, arms, hands, vocal, respiratory, and alimentary systems. Symptoms must be present before age 18, which may wax and wane into adulthood in forms of eyebrow raising, grimacing, winking, or nose-wrinkling.

Genetic factors include 50% of individuals with tic (Tourette's) disorders who also have ADHD and 40% have OCD. Genetic linkage includes multiple familial vulnerabilities associated with the dopamine receptor and transporter genes, androadrenergic and serotonergic genes. Inquiry in new research includes the possibility that there is an autoimmune process with consideration of individuals inflicted by beta-streptococcal infection. Although evidence supporting the PANDAS hypothesis remains limited, one prospective study found that a subset of children with tic disorders and obsessive-compulsive symptoms experienced symptom exacerbations temporally associated with group A beta-hemolytic streptococcal infections (Kurlan et al. 2008).

**Case Study Complexities and Questions**

Erik is an 18-year-old diagnosed with Tourette's syndrome by age 7. Erik's initial symptoms were throat clearing and blinking and evolved with audible vocalizations with motor movements of shrugging his shoulders up. When Erik's anxiety was increased, he had difficulty in school although he was planning to get into college after his senior year of high school.

Assessment: What are his symptoms and for how long have they been present?

What differential diagnosis may account for this?

Treatment: After using a Comprehensive Behavioral Intervention for Tics (CBIT) and school accommodations have been made, permitting extra time to take breaks on tests, and have a longer amount of time on testing, what other treatments is evidence-based pharmacology treatment?

1. Questions (answers in the back of the chapter).
2. What are his symptoms and for how long have they been present?
3. What differential diagnosis may account for this?
4. What evidence-based pharmacology treatment is for tics (Tourette's Disorder)?

### 8.4.1 Evidence-Based Treatment

The first steps in treatment are psychoeducation about the disorder aimed at understanding the variety or cluster of tics or vocalizations. For consideration after the first intervention is to assess and determine whether the Tourette tics are mild, moderate, or severe, using the Premonitory Urge for Tics Scale

(PUTS). The consensus is that both behavioral and pharmacological treatments have been found beneficial. For instance, habit reversal therapy and risperidone have been found beneficial (Borland..). There has been compelling evidence that the dopamine system is involved and is thought that dopamine agonists work best such as haloperidol, pimozide, and fluphenazine, which aimed at suppressing tics and dopaminergic activity. Psychostimulants such as methylphenidate or amphetamine medications increase and worsens tics (Stahl 2024) There is ongoing research related to antipsychotic medications treatment for tics (Tourette's disorder).

## 8.5 Anxiety Disorders

### 8.5.1 Generalized Anxiety Disorder (GAD)

**Case Study**

Sonia is a 33-year-old female, who presents to the clinic with a chief compliant of "difficulty concentrating and persistent restlessness." The history of presenting illness (HPI) includes symptoms started by age 21, worsened with college and job stressors, while chronic anxiety and racing worrying thoughts about job and college performance consumed her thought patterns. She experienced associated symptoms of intermittent irritable mood and difficulty with concentration.

Mental status examination (MSE) included alert, oriented, with mood reported as anxious, affect anxious, mildly anxious vacillating fast speech with thought process as abstract, linear, thought content with anxious and worrying consumptive thoughts and presents easily fatigued. No evidence of inattention or poor concentration.

**Diagnostic Considerations**

ADHD usual onset is childhood, attention that is consistently poor, inattentiveness during the interview, and distractibility. GAD usually presents in early adulthood, ruminating worrying thoughts, and worsens with stressors and common fear of failure.

**Evidence-Based Treatment for GAD**

Cognitive behavioral therapy has benefits associated with the treatment of GAD. This is aimed at treating unrealistic anxiety and/or fears by challenging them, coping skills with uncertainty and reality of likeliness of consequences that trigger the anxious emotions (InformedHealth.org 2024). Psychopharmacology options include selective serotonin reuptake inhibitors (SSRI), serotonin norepinephrine reuptake inhibitors (SNRI), and non-benzodiazepine medication such as buspirone or hydroxyzine or short term benzodiazepines and pregabalin (Strawn et al. 2025; Stahl 2024).

### 8.5.2 Obsessive-Compulsive Disorder (OCD)

**Case Study**

Michael is a 29-year-old software engineer who presents to the clinic with the chief complaint of "intrusive inattention and distracted while at work and home." The history of presenting illness (HPI) includes symptoms of intrusive distractions throughout his workday which makes him feel inept. The symptoms have been present for 3 years; he has been feeling like he has left the cooking stove on or the windows open because of his distracted intrusive thoughts. He is unable to leave his home for work for at least 1 hour after rechecking the stove, windows, and door locks. He feels he has ADHD.

Mental status examination included alert, oriented, mood reported as anxious, affect mildly anxious, with normal speech, thought process abstract, linear, thought content with intrusive obsessive thoughts, and behaviors of rechecking appliances and windows and doors at home and at work. No evidence of delusions or psychosis. No evidence of inattention or poor concentration.

**Diagnostic Considerations**

Obsessions are intrusive thoughts, behaviors, and urges that cause internal distress. Anxiety is the discomfort felt from fear of not attending to the obsessive behavior. The compulsion is a repetitive compulsory behavior that lasts at least 1-hour or longer daily for at least 6 months and is distressing in daily life functioning. Asking the client if the OCD behavior when completed, such as checking the door three times, "Does this relieve the anxious thoughts?"

**Evidence-Based Treatment**

First-line treatment includes cognitive behavioral therapy with exposure and response prevention (Swierkosz-Lenart et al. 2023). When choosing a select serotonin reuptake inhibitor (SSRI) as an anti-obsessive pharmacologic treatment, higher dosages may be required. For example, starting dose for escitalopram is 20 mg, and for anti-obsessive treatment, dosages have been increased to 40 mg daily. Other pharmacologic treatments are fluoxetine ranging from 60 to 80 mg daily, fluvoxamine up to 300 mg daily, paroxetine 40 to 60 mg daily, and sertraline 200 mg daily (Stahl 2024; Swierkosz-Lenart et al. 2023).

Second-line pharmacologic treatment includes clomipramine a tricyclic since only 50% of those with OCD respond to SSRI and serotonin norepinephrine reuptake inhibitors (SNRI) such as venlafaxine. In addition, other adjunctive medication treatment options include low-dose antipsychotic or antidopaminergic medications such as risperidone or aripiprazole, which has been found to be effective (Stahl 2024; Swierkosz-Lenart et al. 2023). Interestingly in treatment-resistant OCD, low-frequency repetitive transcranial magnetic stimulation (rTMS) has demonstrated efficacy for those individuals who have failed at least two different treatment modalities (Swierkosz-Lenart et al. 2023).

## 8.6 Mood Disorders

### 8.6.1 Major Depressive Disorder (MDD), Unipolar Depression

**Case Study**
Bella is a 40-year-old female who works as a registered nurse, with the chief complaint of "poor concentration and inattention and feels she may have ADHD." The history of presenting illness (HPI) includes symptoms of poor concentration, anergia, amotivation, poor sleep, overeating, stress eating, and feeling like there is no passion for the hobbies or work she does any longer lasting for 2 months. The symptoms include inability to have enough energy to leave her home for work and the feeling that this is a concentration issue and thinking she has ADHD.

Mental status examination includes alert, oriented, good eye contact, mood reported is sad, affect mildly blunted, with normal to impoverished speech, thought process abstract, although poor concentration, mostly abstract and linear, thought content with anhedonic behaviors at home and at work. No evidence of delusions or psychosis. No evidence of poor concentration. Insight and judgment are fair.

**Diagnostic Considerations**
Mood disorders, including unipolar depression or major depressive disorder, in the DSM-5 criterion includes poor concentration. The practitioner will consider implementing a Patient Health Questionnaire (PHQ-9) and a Mood Disorder Questionnaire (MDQ) to differentiate between unipolar and bipolar depression. Standardized screening and subjective and objective assessments are the gold standard. According to a recent study, Waheed et al. (2024), despite the opportunity to use a PHQ-9 or MDQ, in addition to an assessment in an integrated primary care setting, 63% physicians were willing to use a PHQ-9 and 32% the MDQ and was not statistically significant in the management of these disorders (Waheed et al. 2024). It is imperative to diagnose unipolar depression or MDD, versus bipolar depression, or bipolar 1 disorder, because the treatment and management are different according to the clinical guidelines.

**Evidence-Based Treatment**
First-line psychotherapy treatment includes cognitive behavioral therapy (CBT) or interpersonal therapy (IPT) (Wheeler 2022). After remitting symptoms, mindfulness-based cognitive therapy can be used for maintenance (Wheeler 2022). First-line psychopharmacology gold-standard treatment is a selective serotonin reuptake inhibitor (SSRI) or selective serotonin norepinephrine reuptake inhibitor (SNRI) (Stahl 2024).

Second-line psychopharmacology therapy includes either tricyclics (TCAs), monoamine oxidase inhibitors (MAOIs), or other antidepressant medications grouped as "others (Karrouri et al. 2021)." For treatment-resistant major depressive disorder, N-methyl-D-aspartate such as esketamine is approved by the US Food and Drug Administration (Karrouri et al. 2021). Each treatment is a patient-by-patient individualized approach.

**Risk Factors**

Risk factors for untreated major depressive disorder include potential suicidal behavior, ideations, or actions. In addition, comorbid anxiety disorders are highly prevalent (Rey-Casserly et al. 2019). On caveat, since MDD is a pressing global mental health disability, potential psychostimulant augmentation of antidepressant medications may alleviate depressive symptoms although risks include potential for dependence or abuse of psychostimulant mediation. Other challenges are comorbid learning disorders or substance use disorders, which require an integrated approach to assessment and treatment decision-making (Janela et al. 2023).

### 8.6.2 Bipolar Affective Disorder (BD)

**Case Study**

Josia is a 29-year-old female who presents for the chief complaint of "poor focus and concentration," and feels she "may have ADHD." The history of presenting illness (HPI) includes symptoms of poor focus and concentration, distractibility, and inability to keep a job which has been on and off for the past year.

**Mental Status Examination**

Mental status examination includes alert, oriented, good eye contact, mood is anxious to labile, affect is expansive, with normal to pressured speech, racing thought process, and derailment on topics in context. Insomnia 4 to 7 nights in a row, phoning people in the middle of the night with mild evidence of delusions of grandeur. No evidence of inattention when asking questions. Insight and judgment are limited to fair.

**Diagnostic Considerations**

Considering an assessment for adult ADHD with a Self-Report Scale (ASRS-vI.I) Symptom Checklist may be clinically indicated, although the clinician must keep the patient outcome and initial diagnosis. First assessment with a MDQ would be most beneficial if suspicion for mania is present. Distractibility is one of the criteria for bipolar 1 affective disorder, and assessment of a MSE on an initial psychiatric interview and diagnostic decision-making would indicate the first diagnosis would be a bipolar 1 disorder secondary to insomnia, racing thoughts, and derailment. The treatment for bipolar 1 disorder and ADHD is different. Once bipolar 1 affective disorder is diagnosed, then the episode and phase such as depressive, hypomania, mania or mixed episode, then treatment could be selected. There is comorbid ADHD and bipolar 1 affective disorder, and lowering the risk for mania is the most important because of safety, and risk for bipolar 1 with psychotic features could be induced (Nierenberg et al. 2023).

**Evidence-Based Treatment**

The gold-standard treatment for bipolar 1 disorder is to select the correct treatment for the current episode. First-line psychopharmacology includes mood stabilizers such as lithium and valproate, lamotrigine, atypical antipsychotic medications such as aripiprazole, lurasidone, cariprazine, and quetiapine (Nierenberg et al. 2023).

**Risk Factors**
Medications that can induce mania include SSRI, SNRI, TCAs, and psychostimulants (Stahl 2024). Approximately up to 20% of adults with bipolar affective disorder complete suicide, and early treatment is imperative (Nierenberg et al. 2023). Once the bipolar affective disorder is successfully treated, then the assessment for comorbid ADHD can be conducted. In addition, it is very successful to treat comorbid ADHD in bipolar affective disorder once the individual is stabilized with a balanced mood.

## 8.7 Impulse Disorders

### 8.7.1 Impulse Control Disorders (ICD)

While impulse control disorders such as conduct disorder or oppositional defiant disorder are diagnosed in children and adolescents, impulse control disorder has overlapping symptoms of attention-deficit disorder or attention-deficit hyperactivity disorder (APA 2025a). Impulse control disorders is an umbrella term for oppositional defiant disorder, conduct disorder, intermittent explosive disorder, and other specified disruptive, impulse-control, and conduct disorders (DSM-5, TR 2022).

**Case Study**
Nicolas is a 21-year-old male who presents for the chief complaint of "impulsivity and concentration problems" and feels he "may have ADHD." The history of presenting illness (HPI) includes symptoms of impulsive thoughts, disruptive behaviors, and poor concentration that occur without any triggering event for the past several years. He has been unable to keep a job or engage in a healthy long-term relationship for more than 1 week.

**Mental Status Examination**
Mental status examination includes alert, oriented, good eye contact, mood is labile to irritable, affect is labile, with normal speech, racing thoughts related to psychostimulant medication. Though content is devoid of psychosis, mania or paranoia in context. No evidence of inattention when asking questions. Insight and judgment are limited to fair.

**Diagnostic Considerations**
Initial assessment questions should include "Have you ever lost control of your temper with law enforcement or people of authority (Zimmerman 2024)?" If yes, "Describe the instance (Zimmerman 2024)." Identify patterns of impulsivity and the surrounding circumstances and if there was any verbal or property damage. Impulsivity alone is just one potential qualifier for ADHD, although rapid outbursts

of angry moods, explosive behaviors, and poor job or interpersonal relationship issues will present for an impulse control disorder (DSM-5, TR 2022).

**Evidence-Based Treatment**

After correct diagnosis has been made, gold-standard treatment is cognitive behavioral therapy (Wheeler 2022). Psychotherapy is aimed at reducing angry impulses while developing relaxing coping skills and mechanisms (DSM-5, TR 2022). Challenging behaviors may include issues with self-control and aggression, and safety is the number one issue that should be addressed that aligns with societal norms. Medications may be targeted to treat the symptoms associated with ICD.

**Risk Factors**

Co-occurring substance use disorders, legal issues, or life-altering events are a risk for individuals with impulse control disorders. It is estimated 2.5% of individuals with ICD have adult ADHD which does lead to disruptive and impulsive behavior (APA 2025a). Many adults may not realize they have ADHD, and early assessment, diagnosis, and treatment are imperative.

## 8.8 Societal and Cultural Stigma of ADHD Comorbidities in Adulthood

### 8.8.1 The Need for Clinical Services for the Treatment of ADHD Comorbidities

**Nosological Considerations**

Nosological patterns include a higher order of traits that include extraversion, neuroticism, conscientiousness, and constraint (Nigg et al. 2020). Cognitive heterogenicity in ADHD involve the symptoms of attention, extraversion, activation in performance, interest in activities, and long-standing reward-based mechanisms (Nigg et al. 2020). In adults, longitudinal studies have been conducted following the child or adolescent into adulthood, and the finding consensus is predominant symptoms of cognitive control related to inattention although may be stimuli driven (Nigg et al. 2020). Temperament in adults with ADHD include from mild-stable to stable or irritability which may be stable versus unstable related to character, personality, treatment of comorbidities, or risks of emotional dysregulation (Nigg et al. 2020). Considerations should include identifying the outcome versus risk of treatment of ADHD comorbidities that have significance in targeted therapies that include organizational skill training and standards of activities of daily living with the focus on positive outcomes.

## 8.9 Clinical Challenges and Real-Life Impairments Related to ADHD Comorbidities

Accurate and early diagnosis of ADHD is beneficial as many co-occurring mental disorders may cause impairment at work, at home, or in college environments. In 1973, the Americans with Disabilities Act (ADA) made it possible for institutions to not be eligible for federal funding reducing risks for discrimination and opening doors for reasonable accommodations (APA 2025b). In addition, a comprehensive psychiatric evaluation may reveal the diagnosis of ADHD, although it is best practice to assess past, current, and medical conditions to accurately diagnose ADHD aimed at reducing real-life impairments.

## 8.10 Conclusion: ADHD Comorbidities Treatment Improves Cognitive Processing

Let us analyze the core deficits to discern the comorbidities aimed at gold-standard evidence-based treatments. Symptoms of inattention and impulsivity should warrant a complete psychiatric evaluation assessment that includes a medical history, evidence-based screening tools, and integrative collaboration with neuropsychological testing. Ruling out other comorbid conditions is very important to diagnose ADHD and other mental health disorders. Cognitive processing may have different levels of severity, and collateral information from family members may be a prudent piece of the diagnostic process.

## References

American Psychiatric Association. What are disruptive, impulse control and conduct disorders? 2025a. https://www.psychiatry.org/patients-families/disruptive-impulse-control-and-conduct-disorders/what-are-disruptive-impulse-control-and-conduct.

American Psychiatric Association. What is ADHD. 2025b. Psychiatry.org. https://www.psychiatry.org/patients-families/adhd/what-is-adhd#section_7.

Autism and developmental disabilities monitoring (ADDM) network. The Centers for Disease Control. 2025. https://www.cdc.gov/autism/addm-network/index.html.

Borland R, Verduin ML, Ruiz P. Kaplan & Sadock's synopsis of psychiatry. 12th ed. PA: Wolters Kluwer; 2022. ISBN-13: 978-1975145569

Diagnostic and Statistical Manual of Mental Disorders (DSM-5-TR). 2022. American Psychiatric Association. Washington, D. C.

Janela D, Jerónimo J, Rema J, et al. Psychostimulant augmentation of antidepressant therapy in depression: a systematic review and meta-analysis. Curr Treat Options Psychiatry. 2023;10:492–510. https://doi.org/10.1007/s40501-023-00307-4.

Gnanavel S, Sharma P, Kaushal P, Hussain S. Attention deficit hyperactivity disorder and comorbidity: a review of literature. World J Clin Cases. 2019;7(17):2420–6. https://doi.org/10.12998/wjcc.v7.i17.2420.

InformedHealth.org [Internet]. Generalized anxiety disorder: Learn More—Treatment options for generalized anxiety disorder. [Updated 2024 Apr 16]. Cologne, Germany: Institute for Quality and Efficiency in Health Care (IQWiG) 2024. Available from: https://www.ncbi.nlm.nih.gov/books/NBK279594/.

Karrouri R, Hammani Z, Benjelloun R, Otheman Y. Major depressive disorder: validated treatments and future challenges. World J Clin Cases. 2021;9(31):9350–67. https://pmc.ncbi.nlm.nih.gov/articles/PMC8610877/#sec2.

Kurlan R, Johnson D, Kaplan EL, Tourette Syndrome Study Group. Streptococcal infection and exacerbations of childhood tics and obsessive-compulsive symptoms: A prospective blinded cohort study. Pediatrics, 2008;121(6):1188–97. https://doi.org/10.1542/peds.2007-2657.

Nierenberg, A. A., Agustini, B, Köhler-Forsberg, O., et al. (2023). Diagnosis and treatment of bipolar disorder: a review. JAMA 2023;330(14):1370–1380. doi:https://doi.org/10.1001/jama.2023.18588.

Nigg JT, Karalunas SL, Feczko E, Fair DA. Toward a revised nosology for attention-deficit/hyperactivity disorder heterogeneity. Biol Psychiatry Cognit Neurosci Neuroimaging. 2020;5(8):726–37. https://doi.org/10.1016/j.bpsc.2020.02.005.

Ohnishi T, Kobayashi H, Yajima T, Koyama T, Noguchi K. Psychiatric comorbidities in adult attention-deficit/hyperactivity disorder: prevalence and patterns in the routine clinical setting. Innov Clin Neurosci. 2019;16(9–10):11–6.

Rey-Casserly C, McGuinn L, Lavin A, Committee on psychosocial aspects of child and family health, section on developmental and behavioral pediatrics. School-aged children who are not progressing academically: considerations for pediatricians. Pediatrics. 2019;144(4):e20192520. https://doi.org/10.1542/peds.2019-2520.

Sadock BJ, Sadock VA, Ruiz P. Kaplan & Sadock's comprehensive textbook of psychiatry. 9th ed. Lippincott, Williams, & Williams; 2009.

Stahl S. Stahl's essential psychopharmacology: prescriber's guide. 8th ed. Cambridge: UK; 2024. ISBN-13: 978-1009464758.

Strawn JR, Mills JA, Ramsey LB, et al. Electronically monitored antidepressant adherence in adolescents with anxiety disorders: A pilot study. J Child Adolesc Psychopharmacol. 2025;35(3). https://doi.org/10.1089/cap.2024.0102.

Swierkosz-Lenart K, Dos Santos JFA, Elowe J, Clair AH, Bally JF, Riquier F, Bloch J, Draganski B, Clerc MT, Pozuelo Moyano B, Von Gunten A, Mallet L. Therapies for obsessive-compulsive disorder: Current state of the art and perspectives for approaching treatment-resistant patients. Front Psych. 2023;14(16):1065812. https://doi.org/10.3389/fpsyt.2023.1065812. PMID: 36873207; PMCID: PMC9978117

Waheed A, Afridi AK, Rana M, Arif M, Barrera T, Patel F, Khan MN, Azhar E. Knowledge and behavior of primary care physicians regarding utilization of standardized tools in screening and assessment of anxiety, depression, and mood disorders at a large integrated health system. J Prim Care Community Health. 2024;15:21501319231224711. https://doi.org/10.1177/21501319231224711.

Wheeler K. Psychotherapy for the advanced practice nurse: a how-to guide for evidence-based practice. 3rd ed; 2022. ISBN-13: 978-0826193797

## *Resources*

Zimmerman M. Interview guide for evaluating DSM-TR psychiatric disorders and the mental status examination. E. Greenwich: Psych Products Press; 2024.

# 9 Navigating the Legal and Regulatory Issues for the Use of Telehealth in ADHD Management

Mary Ann Camilleri

## 9.1 Telehealth: State of Play

Telehealth, used for over 30 years in rural communities, surged during the COVID-19 pandemic (Hyder and Razzak 2020). Relaxation of regulatory reimbursement requirements enabled the penetration of telehealth nationally as providers and healthcare systems recognized it as a viable care model to improve access and expand capacity. With the rapid deployment of telehealth applications and expanding digital infrastructure, the National Academy of Medicine's vision of transforming healthcare through technology appeared to be strongly rooted (IOM 2001, p.16). Although reduced demand for telehealth post-pandemic, regulatory uncertainties, and other barriers have tempered expansion in some arenas, the demand for telehealth use for low acuity behavioral health remains strong (Patton et al. 2024).

Telehealth offers a robust alternative to in-person care for the ongoing management of attention-deficit/hyperactivity disorder (ADHD). About 15.5 million or 6% of adults in the United States have been diagnosed with ADHD, half of whom use telehealth for ADHD-related services (Staley et al. 2023). As of 2022, 6.5 million or 10.5% of US children, ages 3–17, had a current diagnosis of ADHD (Danielson et al. 2024). In a cross-sectional study of 112 parents and caregivers of children with ADHD who had experience with telehealth, the majority (81.3%) were willing to use telehealth for future appointments for their children (Galvin et al. 2024). Perceived benefits of telehealth included convenience, time savings, cost reductions, improving the family's routines, and offering a useful option for children who are not comfortable with in-person visits (Galvin et al. 2024). Reduced waiting times enabling more timely access, avoiding stigma, protection from seasonal infectious diseases, and improved access for those with geographical, transportation, and

M. A. Camilleri (✉)
School of Nursing, Yale University, New Haven, CT, USA
e-mail: maryann.camilleri@yale.edu

© The Author(s), under exclusive license to Springer Nature Switzerland AG 2026
R. J. Kennedy (ed.), *Assessment & Treatment of ADHD*,
https://doi.org/10.1007/978-3-032-24171-9_9

mobility limitations are identified benefits (Galvin et al. 2024; Uniform Telehealth Act (UTA) 2022a, 2022b). A systematic review of research articles on the use of telehealth for mental health services found it to be effective as an adjunct to conventional in-person care (Langarizadeh et al. 2017).

Despite its advantages, barriers to the advancement of telehealth are the proliferation of regulation and uncertainty of the regulatory environment. The practitioner wishing to provide this desired service is faced with a conglomeration of federal and state laws and regulations, many of which are in flux, requiring constant diligence for compliance and strategic decision-making. While not a compendium of every law and regulation applicable to telehealth practice, this chapter provides guidance to practitioners about key requirements and developing issues in a rapidly changing regulatory environment.

## 9.2 Overview of the US Regulatory Scheme for Telehealth

Statutory law in all 50 states, Washington D.C. and Puerto Rico define telehealth and address professional requirements such as licensure pathways, informed consent, privacy and confidentiality, prescribing, scope of practice, standard of care, and medical malpractice insurance requirements (Center for Connected Health Policy (CCHP)). A plethora of reimbursement regulations of government and commercial payors include varying technology and practice restrictions, adding further complexity. Some of the federal restrictions, removed or relaxed during the pandemic, have survived only through a sequence of temporary extensions, while few have become permanent, creating continued regulatory uncertainty for practitioners.

Other federal laws implicated in the provision of telehealth professional services include the Privacy and Security rules of the Health Insurance Portability and Accountability Act of 1996 (HIPAA), (P.L. 104-191), Health information Technology for Economic and Clinical Act of 2009 (HITECH), (P.L. 111-5), the Ryan Haight Online Pharmacy and Consumer Protection Act of 2008, (P.L. 425). The Health and Human Services Office of Civil Rights and Department of Justice Civil Rights Division issued guidance pertaining to legal obligations of providers to provide equal access of telehealth and nondiscrimination based on disability, race, color, national, origin, sex, and limited language proficiency (OCR and DOJ 2025). Health and Human Services and the Office of Inspector General issued a Special Telehealth Fraud Alert which addresses business practices of companies and individual practitioners (OIG 2022). While advisory and nonbinding, the fraud alert and subsequent enforcement actions are instructive to practitioners in avoiding criminal liability for fraud and abuse in telehealth services. Layered upon the legal and regulatory requirements are the provisions of commercial payor contracts, often following Medicare rules, which further bind the participating telehealth practitioner through their contractual obligations.

### 9.2.1 The Uniform Telehealth Act

Recognizing the benefits of telehealth, the Uniform Law Commission is, a nonpartisan advisory body working for uniformity of state laws, established a multi-state collaborative of legal experts, regulators, and stakeholders culminating in the Uniform Telehealth Act of 2022. The Act defines telehealth, authorizes a practitioner-patient relationship established through telehealth, permits practitioners to provide telehealth services within their scope of practice subject to state and federal law, requires adherence to certain practice standards, and provides licensure pathways for out-of-state practitioners to treat patients using telehealth (Uniform Telehealth Act (UTA) 2022a, 2022b). While the act is advisory only, it provides uniform guidance to state and federal regulators with a secondary goal of expanding telehealth services.

The Uniform Telehealth Act defines telehealth as "use of synchronous or asynchronous telecommunication technology by a practitioner to provide health care to a patient at a different physical location than the practitioner" (UTA, §2 (12)). Healthcare is defined as "care, treatment, or a service or procedure, to maintain, monitor, diagnose, or otherwise affect an individual's physical or mental illness, injury, or condition" (UTA, §2 (3)). Telehealth services refer to healthcare provided through telehealth (UTA, §2 (13)). This broad definition encompasses both real-time communication such as virtual visits using visual and audio, or audio only, and delayed communication methods for healthcare delivery such as chat-based interactions, email, or text. Remote patient monitoring (RPM) and other technology-enabled modalities may also be included. Definitions of telehealth vary by state statute, state licensing boards, and federal statute and by payor resulting in differences in the covered technologies and services that will be reimbursed.

Currently, only the state of Washington and the District of Columbia have adopted the Uniform Telehealth Act in its entirety, while several others may have adopted one or more provisions of the Act in whole or in part.[20] As a result of this lack of standardization, it is incumbent upon the telehealth practitioner to know the rules of each state applicable to their practice.

## 9.3 Professional Requirements

There are six common legal concepts that are threshold practice issues for the practitioner who desires to offer telehealth services: licensure, prescribing authority, informed consent, privacy and confidentiality, standards of care and professional liability insurance (see Appendix 9.1. Checklist of Professional Requirements for Telehealth Practice). An overview of each concept describes the fundamental legal obligations; however, specific state requirements vary. The Center for Connected Policy provides an interactive policy tool for state-to state rules and comparisons at https://www.cchpca.org/all-telehealth-policies/ (CCHP 2025).

### 9.3.1 Licensure

Professional licensure is the province of each state under its power to protect the public welfare, health, and safety of its residents. In addition to professional licensing statutes, many state telehealth statutes specify which types of licensed healthcare professionals can provide telehealth services. Categories of providers, such as physicians, advanced practice nurses, psychologists, and others, may be authorized to deliver telehealth care. When the patient resides and the provider practices in the same state, the license is issued by a single state.

The situation becomes more complex when the practitioner seeks to provide health services to a patient who is located or resides in a different state (originating site) from in the provider's own state of practice (distant site). Because the telehealth service is deemed to be provided in the state of the patient's is location, issues of cross-state practice and multi-state licensing arise. Given that each state has its own licensing authority, the provider is required to be fully licensed in both the distant and originating states. Challenges to the geographic restrictions of telehealth due to the burdens of licensing in multiple states have been made in the courts, especially regarding telehealth access to nationally recognized specialists for consultation.

The Pacific Legal Foundation has filed two such cases, Shannon McDonald, M.D., et al. v. Otto Sabando (USDC, D. New Jersey, 2023) and Sean McBride et al., v. Lawson (E.D.Calif., 2024) alleging New Jersey and California physician licensure laws are violations of the United States Constitution Commerce Clause, Privileges and Immunities Clause and First Amendment. The Court dismissed each constitutional claim. Regarding the First Amendment the court observed that although not alleged, there is a "viable distinction between speech that is 'treatment' and speech that is not" (McBride, p17) seemingly opening a door for a litigant to reframe the First Amendment challenge. In Elizabeth Brokamp v. Leticia James (U.S.C.A. 2nd Cir., 2023) the plaintiff Elizabeth Brokamp a licensed counselor in Virginia provided talk therapy via telehealth in New York. Brokamp brought a First Amendment claim of content -based restriction on speech, alleging that professional counselors do not prescribe or perform medical procedures. The court dismissed the case finding that the licensure law for mental health counselors does not depend on what the counselor discusses so it is not a content restriction and no injury resulted from the licensure requirement. The court's decision was affirmed on appeal to the Second Circuit. The US Supreme Court declined to hear the case. The court decisions demonstrate that federal courts are inclined to defer to the state's regulation of professional licensure, forecasting a dim outlook for the success of constitutional challenges to individual state authority for licensure and telehealth.

### 9.3.2 Alternate Pathways to Licensure

Alternative pathways to streamline multi-state licensure for telehealth and other providers are emerging. During the COVID pandemic, several states established special telehealth registration or specific telehealth licensure to expedite the process for physicians and other providers who hold licensure in their home state (NCSL). After COVID, 22 states offer telehealth registration, though requirements vary among the states. Typically, the provider must hold a valid unrestricted license in another state without any past disciplinary proceedings, must pay an annual registration fee with the designated state agency, and must provide evidence of professional liability insurance. In some states the telehealth registered provider is prohibited from having an office or providing in-person care in the state (CCHP 2025). The in-state office restriction may conflict when a federal, state, or private payer requires in-person visit at some interval in the treatment. Logistical and cost issues are created for the out-of-state practitioner who will need to contract with an in-state provider for the in-person service.

Some states, such as South Carolina, limit registration to specific categories of professionals, such as behavioral health professionals. Other states limit to certain types of services such as short-term follow-up care for purposes of continuity, emergency care, or virtual consultations with a specialist (CCHP 2025).

Interstate compacts have gained the interest of professionals and professional associations to ensure proper credentialing of professionals who either move to another state or wish to practice across state lines. While interstate compacts are well developed, for some types of professionals, such as physicians, and other professions, such as psychologists, they have been slow to progress. The Interstate Medical Licensure Compact offers expedited medical licensing in 43 participating states, Washington D.C. and Guam (CompHealth 2026).

The compact presents a streamlined process for a practitioner licensed in a state designated as a state of principle licensure to obtain licensure in other participating states, without changing the state's practice act (CompHealth 2026). The telehealth physician remains subject to the jurisdiction of the medical board of the state where the patient is located. Participation in the compact is dependent on the approval of the professional licensing board and the passage of compact legislation in the state. For example, as of June 2025, the State of California has not joined the Association of State and Provincial Psychology Board's Interjurisdictional Compact (PSYPACT), though it has been formally approved in 41 states and adopted in 39 states. The ongoing debate centers on the California licensing board's concerns *of sharing licensee information, the entry of competitive psychologists to practice in the state, the stringent requirements of the compact that existing California psychologists may not meet, and surrendering regulatory control to the Psychology Interjurisdictional Compact Commission, a non-governmental out-of-state entity (Maheu 2024). Others argue that the failure to enter PSYPACT ignores the opioid crisis, increasing mental health needs, and access issues in rural areas.* The Center for Connected Healthcare Policy (CCHP) provides an interactive tool tracking the progress of the Interstate Medical Licensure Compact and other compacts (cchpca.org). The National Council for the State Boards of Nursing provides information regarding the Nurse Compact with currently 43 states participating. The APRN Compact is still emerging, in part due to the lack of consensus on state scope of practice requirements for APRNs and specialties (ncsbn.org).

### 9.3.3 Prescribing

The Federal Drug Enforcement Agency (DEA) regulates the prescription of controlled substances via telehealth under the Controlled Substances Act (CSA), Title II of the Comprehensive Drug Abuse Prevention and Control Act of 1970, and the Ryan Haight Online Pharmacy Consumer Protection Act of 2008 (Ryan Haight Act) (Ortiz and Preuss 2024). Practitioners prescribing controlled substances must be registered with the federal DEA in the state where the patient is located (21 CFR §1301.12). The Ryan Haight Act requires a provider to examine a person at least once before prescribing a controlled substance (21 USC §829e). Accordingly, at least one in-person evaluation would have to occur before an authorized provider could prescribe a stimulant via telehealth for a patient with ADHD. During the COVID-19 pandemic, an exception allowed DEA-registered practitioners authorized to prescribe schedule II–V controlled substances via telemedicine, to prescribe without an initial in-person evaluation. Through a series of extensions, the exception was temporarily extended through December 31, 2027 (21 CFR §1307.41).

In January 2025, the DEA released the proposed rule, Special Registrations for Telemedicine, and Limited State Telemedicine Registrations (90 FR 6541). The purpose is to provide more comprehensive patient access to controlled substance medications via telemedicine while mitigating the risks of diversion through heightened prescription, recordkeeping, and reporting requirements. The proposed rule provides for three categories of Special Registrations for Telemedicine Prescribing:

1. Telemedicine Prescribing Authorization: Authorizes qualified *clinician practitioners* (e.g., physicians and mid-level practitioners) to prescribe Schedule III–V controlled substances.
2. Advanced Telemedicine Prescribing Authorization: Authorizes qualified *specialized clinician practitioners* (e.g., psychiatrists and hospice care physicians) to prescribe Schedule II–V controlled substances.
3. Telemedicine Platform Registration Authorization: Authorizes online telemedicine platforms, in their capacity as *platform practitioners*, to dispense Schedule II–V controlled substances.

The Special Registrants in all three categories would be required to maintain a State Telemedicine Registration in every state in which the special registrant treats patients. The State Telemedicine Registration is an ancillary requirement issued by the DEA, not by the state (90 FR 6541). In the interest of streamlining the application process, a new form will be issued and would be valid for 3 years. Each special registrant will be assigned identifying numbers for their Special and State Telemedicine Registrations. Special registrants must comply with stricter requirements for prescribing including using their special registration numbers, electronic prescribing for controlled substances (EPCS) after verification of the patient's identity, and a nationwide Prescription Drug Monitoring Program Check (PDMP).

In the Telemedicine Listening Sessions for the proposed rule, speakers raised an important accessibility issue for patients with ADHD who benefit from the widely accepted treatment with Adderall or other Schedule II controlled substance (90 CFR 6550). The new rule requires that only a special registrant of the second type, Advanced Telemedicine Prescribing Authorization, may prescribe a Schedule II controlled drug. The proposed rule, as introduced, seems to indicate that only a board-certified psychiatrist would be authorized to treat and diagnose ADHD. If so, this rule would be more stringent than the COVID flexibility extended until December 31, 2027, creating access issues and hardships for patients rather than presenting a permanent solution. In response the DEA clarified that the application for the Advanced Telemedicine Prescribing Authorization requires the specialized physician or "mid-level practitioner" to be able to demonstrate "specialized training and in-depth knowledge to equip them to make informed decisions regarding the use of Schedule II controlled substances when prescribed remotely to particularly vulnerable patient groups" (90 CFR 6550). The DEA made a distinction between qualifications of physicians or psychiatrists on the one hand, and qualifications of the "midlevel practitioners" such as Advanced Practice Registered Nurses (APRNs) and Physician Assistants (PAs). Specifically, it is sufficient for a physician applicant to provide information on board certification in a specialty, specialized training, or the percentage of the *physician's* overall practice that falls within one of the specialized practices. For *mid-level practitioners, however, the applicant must* be board-certified *under the proposed framework.* Attempting to avoid the delineation of specific criteria for each specialized practitioner, the DEA invited public comment and input on other types of practitioners commonly treating conditions that require the prescribing of Schedule II controlled substances as well as alternate methods of demonstrating training and expertise (90 CFR 6550). The resolution of these issues around qualification for Advanced Telemedicine Prescribing is critical to ensure access to medication treatment, affordability, and convenience for populations with ADHD seeking to continue medication treatment with qualified Advanced Practice Nurses via telemedicine.

In addition to federal prescribing laws, states vary in prescribing authority and scope of practice for types of practitioners. The DEA depends upon state licensing boards to determine whether a practitioner is qualified to dispense, prescribe, or administer controlled substances. These boards also determine which drug categories (schedules) can be dispensed, prescribed, or administered. Both the requirements and the registration process for controlled substance registration vary by state. Prescribers may be required to register with the state prescription monitoring program (PMP or PDMP), so that the state can oversee prescription and dispensing activity and monitor for abuse (see CCHP 2025).

### 9.3.4 Informed Consent for Telehealth

Informed consent is a basic tenant of health law in the United States, a policy recognizing the competent adult patient's autonomy to accept or refuse treatment. The essential elements of informed consent are an explanation of the nature of the treatment, benefits, and risks of the treatment or nontreatment, reasonable alternatives and the risk and benefits of the alternatives, and an assessment of the patient's understanding of the discussion (Shah et al. 2024). The most important distinction is that the informed consent is not a pro forma document but rather a process requiring the patient to have the capacity to consent and to have reached an understanding which the provider assesses and documents.

It is generally accepted that consent is addressed at the time of intake for General Consent to Treatment or, more specifically, prior to proceeding with a procedure or administration of high-risk medications. Similarly, as an emerging modality that may be associated with risks different from in-person interactions, most states require that before proceeding with telehealth, the provider must obtain specific telehealth-informed consent from the patient (CCHP 2025). The American Psychological Association refers to obtaining dual consent, consent for psychological treatment and consent for the use of the technology (APA Telepsychology Guidelines, p.9). At a minimum, this means that the patient consents to the use of telehealth as an acceptable mode of delivering healthcare services including an understanding of the mode of communication used and its limitations. For example, the provider should discuss areas specific to the technology: (a) the nature of services, interventions, and assessments, (b) communication modalities used in these services (e.g., videoconferencing, audio-only telephone calls, chat, patient portals), (c) potential risks and benefits of the service (e.g., technological problems causing interruption and effects on treatment, (d) service limitations, and (e) specific privacy and security measures.

The provider is required to document the consent, usually by a signed consent form through the patient portal. The provider should also briefly document the discussion and confirm the patient's understanding, best accomplished through teach-back method. An interpreter may be required for patients with Limited English Language Proficiency. Consideration should be given to capacity and age. While parents are legally required to consent for a minor, the minor should be included in the discussion to the extent of their ability to understand, particularly when the minor is an adolescent.

AHRQ recommends using the telehealth consent form as a checklist during the discussion (AHRQ 2020). The consent form may have to be modified to address the specific requirements in the state's telehealth statute, administrative agency regulations, or state's Medicaid regulations (Appendix 9.2).

### 9.3.5 Professional Liability and Insurance

With the expansion of telehealth, it is anticipated that professional liability associated with virtual modalities of care are likely to emerge. Concerns have focused on limitations of the virtual clinical environment which may implicate verification of patient identity, inadequate history taking and access to medical records, incomplete visual assessment, ineffective communication, diagnosis, treatment including prescribing, establishing a trusting provider-patient relationship, and continuity of care. Other potential areas include lack of informed consent, as discussed in the previous section, privacy and confidentiality, and data security breaches. While it has been argued that given the nuances of virtual care, a national standard of care for telehealth should be recognized, the majority view contained in state legislation establishes that the provider is held to the same standard of care in virtual care as in-person care (Wolf 2020). It is generally accepted that the patient should have the same expectation of receiving quality care whether it is provided in-person or via a web-based virtual platform or audio-only means. There are exceptions in a minority of states, such as Hawaii, where recommendations given via telemedicine "shall be held to the same standards of appropriate practice as those in traditional physician-patient settings that do not include a face-to face visit" (Hawaii Rev. Stats §453–13). The minority standard allows courts to consider technology's limitations offered on expert witness testimony.

The duty to provide care that complies with the certain standards arises out of the existence of a provider-patient relationship. For legal purposes, a provider-patient relationship is likely formed when the patient has sought care from the provider and the provider and patient agree to that care, without the need for an in-person meeting. Generally, the standard of care is that of a reasonably prudent similar provider under like or similar circumstances. It is only when the duty of care owed to the patient is breached and causes injury and compensable damages that a cause of action for medical malpractice arises. It is incumbent upon the practitioner to be aware of technology-specific risks and to be proactive to prevent and manage those risks.

Risk management guidance is typically obtained from closed claims analysis, either of reported cases reaching a verdict in court, or payments made in settlement by an insurer. As an emerging area of liability, there are insufficient cases to offer such guidance. In a review of the LexisNexis database from October 1, 2018, to November 1, 2018, no reported cases alleging telemedicine malpractice were found (Fogel and Kvedar 2019). The brief study period and the focus on reported cases which excludes cases mediated or settled out of court are limitations of the study. Nevertheless, the need for better competency-based education has been identified (Rowland et al. 2022). In a study of US-based family physicians, while most physicians reported confidence in taking medical histories over telehealth, they had low confidence in managing multiple complaints, language difficulties, prescribing, and logistics such as patient transport (Venditti et al. 2022).

#### 9.3.5.1 Telehealth Practice Guidelines

Increasingly, state licensing boards and other professional organizations are developing telehealth guidelines for the telehealth practitioner. While not a regulatory mandate, professional guidelines can be particularly useful because they address risks associated with both the modality and the specialty. In a lawsuit or licensure proceeding, practice guidelines may be used by an expert witness to establish evidence of standard of care in telehealth practice. The American Psychological Association Guidelines for the Practice of Telepsychology are illustrative.

##### 9.3.5.1.1 Competency of the Practitioner

The APA telepsychology guidelines identify competencies required for the telehealth provider. Three domains of competencies for telehealth practice are (1) Content Competencies, (2) Technical Competencies, and (3) Population Competencies. Content Competencies focus on the type of technology appropriate for specific patient demographic characteristics and symptomology; differences between in-person and telepsychology visits; care considerations or adaptations for history taking, assessment, intervention, establishing rapport; ethical considerations such as informed consent; knowing the legal requirements of cross-state practice; safety planning and practice logistics such as documentation; appropriateness of individual versus group therapy; and documenting. Technical Competencies focus on understanding the technology used in their practice including compliance with regulations, data security, troubleshooting, and being able to communicate troubleshooting methods to others. Population Competencies include the ability to evaluate whether the technology is appropriate and beneficial for the patient, especially special populations such as elderly, children, or individuals with disabilities or comorbidities that require special considerations.

##### 9.3.5.1.2 Patient Considerations

Before proceeding with virtual care, clinical best practice begins with an assessment of the patient, in-person or remote, to consider the patient's appropriateness for using the modality, such as videoconferencing or audio, for ongoing treatment (APA, p.13). Cultural, linguistic and socioeconomic, and other health factors such as psychiatric stability, personal preferences, and the setting should be assessed. For children and adolescents receiving telehealth services for ADHD, co-occurring conditions, such as autism spectrum disorder or severe mental disorders, may present challenges that require accommodations (see Galvin et al. 2024). Similarly, young children with ADHD who experience more severe symptoms, such as distractibility, may be difficult to manage in virtual visits. The adolescent who has concerns about parental presence affecting privacy and confidentiality may require coaching and assistance to negotiate privacy or may prefer in-person visits. For patients with ADHD, physical comorbidities or adverse effects of medication are common and may require augmenting virtual visits with remote physical monitoring devices (Galvin 2024).

#### 9.3.5.1.3 Equitable Access to Telehealth for Special Populations

When completing a patient assessment to determine the patient's appropriateness for virtual care, adaptations and reasonable accommodations to address the patient's needs for accessibility to telehealth services may be required. Federal civil rights laws, such as Section 504 of the Rehabilitation Act, the Americans with Disabilities Act, Section 1557 of the Affordable Care Act (ACA), and Title VI of the Civil Rights Act prohibit discrimination in the basis of disability, race, color, national origin, sex, and age. Federal regulations specifically require that electronic or information technology must be accessible to person with disabilities unless doing so would result in "undue financial and administrative burdens or would fundamentally alter the health program" 45 C.F.R § 92.104(a). The Health and Human Services Office of Civil Rights (OCR) and the Department of Justice issued guidance on nondiscrimination to ensure telehealth access to people with disabilities and Limited English Proficient Persons (LEPP). Some examples of reasonable accommodations include an interpreter or instructions in another language for LEPP, extended visit time for persons with cognitive disabilities, the ability to join a sign language interpreter in videoconferencing for deaf individuals, and enabling screen reader software for a visually impaired person. Detailed guidance with links to FAQs and helpful advisories are contained in the OCR and DOJ's Guidance on Nondiscrimination in Telehealth: Federal Protections to Ensure Accessibility to People with Disabilities and Limited English Proficient Persons at https://www.hhs.gov/sites/default/files/guidance-on-nondiscrimination-in-telehealth.pdf.

#### 9.3.5.1.4 Planning for Continuity of Care

Selection of the right modality with the patient should include a discussion about access to stable broadband, availability of equipment, the patient's competency to use the equipment and the setting, private and free of interruptions and distractions. A plan should be made with the patient in advance to ensure continuity of care, escalation to an in-person visit, change of modality (videoconferencing to audio only) in the event of equipment malfunction, and an emergency or safety care plan (APA, p.17). The preparation for virtual visits through this comprehensive assessment and planning can help to mitigate the risks.

#### 9.3.5.1.5 Virtual Communication and the Therapeutic Relationship

Communication, often challenging in the in-person environment, is more nuanced in the virtual environment. In addition to the factors discussed here, nonverbal cues and subtleties in body language may be more difficult to discern in videoconferencing communication and completely lacking in audio only. A concern in any type of remote behavioral health telemedicine practice is whether the provider could miss important behavioral clues. An area not yet fully studied is whether the loss of nonverbal communication and therapeutic touch in the virtual care environment impairs the telemedicine rapport (Shaver, 526). This is especially important because patient understanding, adherence, and outcomes are known to be heavily influenced by effective clinician communication techniques (Shaver, p.526). Moreover, research has shown that a positive clinician-patient relationship increased patient perception of clinician competence and reduced their intention to bring a malpractice suit when an adverse outcome occurred (Moore 2000).

For these reasons, the clinician must be highly astute to their own communication practices. Recommended techniques for virtual communication include opening by greeting the patient, establishing eye contact by looking into the camera, telling the patient when you need to look away from the screen to document, maintaining a reasonable pace of speech and speaking clearly, actively listening and demonstrating empathy. Using education tools such as teach-back method to ensure the patient's understanding, involving care partners if the patient consents, and always verifying patient's understanding and providing an opportunity for questions before ending the visit are best practices (CNA/NSO 2023). Closing the communication loop through timely follow-ups is essential for care continuity. Especially for the clinician who may provide telehealth coverage via a platform provider and does not have the benefit of a prior therapeutic relationship with the patient, protocols to ensure review and response to testing, continuity of care, and specialty referrals are crucial.

#### 9.3.5.1.6 Professional Liability Insurance: Telehealth Coverage

Risk financing, or professional liability insurance (PLI), is an essential risk management strategy. While PLI cannot protect the clinician from a medical malpractice lawsuit, in the event of a claim, PLI provides the financial and professional resources necessary to effectuate a defense and pay a judgment or settlement. Many states require professional liability insurance for clinician licensing and practicing in the state. In some states, the telehealth statute specifically requires PLI. As telehealth has expanded, many professional liability insurers automatically include coverage for telehealth practice in the standard policy. However, depending on the insurer and clinical specialty, the traditional PLI policy may contain restrictions on coverage or exclusions for telehealth. If a clinician is considering telehealth practice, they should first contact their insurer to determine if there are specific telehealth services that are covered, or excluded, under their existing policy. If telehealth is not already covered, the clinician will need to purchase a telemedicine or telehealth coverage rider to their existing policy. The telemedicine coverage rider provides coverage for medical malpractice claims arising out of the provision of telehealth services. There will likely be an additional application, disclosure, additional coverage limits, and premium requirements. The amount of the premium will be rated based on several factors including the type of practitioner, type of telehealth services, location of services such as whether practicing across state lines and the number of states, and the medical malpractice climate or experience in the state(s) in which you are practicing.

Two important conditions for coverage should be noted. First, on the insurance application, the clinician must fully disclose all telehealth modalities, all states licensed in and practicing in- person or virtually, and their scope of telehealth practice including prescribing specifics. Second, telehealth coverage generally will be conditioned upon proper licensure or registration in the states, including DEA registrations. Most PLI policies contain a provision requiring a continuing duty to disclose. This duty means that if during the term of the policy the clinician plans to practice in other states or expands their scope of practice, they must promptly notify the insurer. Failure to do so can result in a denial of coverage for a specific event. Most denials are due to practicing outside of scope of licensure or intentional acts or criminal liability, such as a government action for fraud and abuse, defamation, or ignoring professional boundaries.

#### 9.3.5.1.7 Other PLI Coverage Issues

Coverage for cybersecurity may require a separate endorsement and additional cost. It typically provides coverage for private actions brought against a clinician or practice for security and privacy breaches or noncompliance with security and privacy regulations. Cybersecurity coverages also insure for customer notification expenses for actual or suspected breach and expenses for management of the clinician's or practice's public image potentially damaged by the breach.

It is not unusual for a clinician involved in an adverse outcome to find themselves defending in two forums, a malpractice case in court and a licensure proceeding. Often the PLI policy only recognizes a complaint filed in court in a malpractice case as a covered risk and, if so, will not provide a defense in a licensure proceeding. Administrative coverage provides for the cost of defense in a state licensing board action, such as negligence or unprofessional conduct. While administrative coverage is often included in an individual policy, it may not be included in the practice employer PLI, so should be purchased individually.

Whether for defense of a lawsuit or an administrative claim, the clinician should look for a policy where coverage for defense costs is provided in addition to the coverage limits for the actual claim, judgment, or settlement. This feature is important to avoid dilution of policy limits by defense costs. When purchasing any PLI policy, the clinician should take care to purchase from a company that is highly rated by one of the top 5 insurer rating agencies. While the rating scales vary by agency, the highest rating of the agency signifies financial stability of the insurer. The clinician should review all coverages, exclusions, and responsibilities of the insured, such as for timely notice to the insurer of a potential or actual claim.

## 9.4 HIPAA Privacy and Security and Telehealth

The Health Insurance Portability and Accountability Act of 1996 (HIPAA), Public Law 104-191, authorized the Secretary of Health and Human Services (HHS) to publicize standards for the electronic exchange, privacy, and security of health information. In addition to HIPPA, its counterpart, the Health Information Technology for Economic and Clinical Health (HITECH) Act of 2009, federal substance use disorder regulations (42 CFR. Part 2), and the Federal Trade Commission's (FTC) Health Breach Notification Rule (16 CFR Part 13) for personal health records (PHRs) also impact how health information and other patient data may be accessed, used, and disclosed. In addition, the clinician must comply with state laws addressing privacy and confidentiality of patient information which may contain more stringent requirements. Since the enactment of HIPAA, numerous regulations attempt to clarify its requirements and to keep pace with the rapid technological shift toward digital healthcare delivery. While this section provides an overview of recent developments in HIPAA and telehealth, the regulatory complexity applying to any service expansion or business arrangement augmenting telehealth should not be underestimated. For example, remote patient monitoring, artificial intelligence applications for analytics, or collaborations with other platform providers such as prescribing services should be embarked upon carefully, not only with technology experts but also in consultation with legal experts specializing in this rapidly growing area of data privacy and security.

### 9.4.1 HIPAA Basics and Key Terms

The HIPAA rules must be applied to secure the privacy of remote patients to the same extent as in-person patients. HIPAA contains rules in three categories:

1. Privacy: Sets national standards for the protection of individually identifiable health information (IIH).
2. Security: Sets national standards for the protecting confidentiality, integrity, and availability of electronic Protected Health Information (PHI).
3. Enforcement: Sets standards for enforcement by DHHS, Office of Civil Rights.

Health Information Technology for Economic and Clinical Health Act of 2009 (HITECH Act) strengthens HIPAA through five additional categories of rules:

1. Clarifies that HIPAA applies to business associates, or vendors such as platform providers, software developers, and other business partners
2. Sets a minimum necessary standard for the use and disclosure of PHI
3. Provides individual rights of access and right to amendment of PHI
4. Provides for breach notification
5. Establishes penalties for violations of HIPAA including treble damages

*Covered entities* include a health plan, healthcare clearinghouse, or healthcare provider who transmits any health information in electronic form in connection with a transaction for which DHHS has set standards. In short, any healthcare provider who bills an insurer for professional services is a covered entity subject to HIPAA.

*Protected Health Information (PHI)* is individually identifiable health information (IIHI) that is transmitted by or maintained in electronic media or transmitted or maintained in any other form or medium (45 CFR 160.103).

*Individually identifiable health information (IIHI)*, a subset of PHI, is information:

- Created or received by a healthcare provider, health plan, employer or healthcare clearinghouse.

- That relates to an individual's past, present, or future physical or mental health or condition, the provision of care to an individual, or the past, present, or future payment for the provision of healthcare to an individual.
- That identifies the individual, or there is a reasonable basis to believe the information could be used to identify the individual (45 CFR 160.103).

HIPAA specifies 18 identifiers of IIHI. Examples are social security numbers, telephone numbers, addresses, web URL, Internet Protocol (IP) address, and finger or voice print. The purpose of the designated identifiers is to allow for the de-identification of health data for research and public health purposes (45 C.F.R §164.514). However, the storage and handling of IIHI by the vendor of purchased applications such as scheduling portals, educational websites, and analytics can present a PHI risk. This IIHI metadata can be linked to PHI and result in unintended disclosures for which the provider would be responsible. For example, if a provider has a teaching module on their website, such as for information about ADHD which the patient accesses from their computer, the patient's URL is IIHI. Storage or analytics data functions embedded in the platform can cause the IIHI to be linked to ADHD diagnosis, becoming PHI. It is imperative for the provider to enter into a Business Associate Agreement with the vendor that contains specific provisions as to how the data including IIHI is accessed, stored, and secured (Giovannini and Watkins 2025).

A *business associate* is a person or entity, other than an employee of the covered entity, that performs certain functions or activities that involve the use or disclosure of protected health information on behalf of, or provides services to, a covered entity. The types of functions or activities that may make a person or entity a business associate include payment or healthcare operations activities, claims processing, or administration; data analysis, processing, or administration; utilization review; quality assurance; billing; benefit management; practice management; accounting; consulting; data aggregation; management; administrative; accreditation; and financial (45 CFR 160.103).

Any subcontractor or vendor that provides services to a provider related to the telehealth platform or augmented digital services is a business associate. A provider is responsible for privacy or security gaps by the business associate that results in an unauthorized disclosure of PHI. Whether or not on the vendor's represents that the platform is verified as HIPAA compliant, a BAA is required for the provider's legal protection. A BAA is a legally binding contract between a covered entity and business associate that details how PHI will be accessed, stored, and protected. The clinician should avoid using services that refuse to provide a BAA, not uncommon among companies who offer unpaid analytic services. The Northeast Telehealth Resource Center provides a downloadable stepwise guide as a quick reference to HIPAA compliance and business associate basic for telehealth providers at https://netrc.org/wp-content/uploads/2015/11/HIPAA-Stepwise-Guide.pdf

### 9.4.2 HIPAA Compliant Telehealth Delivery Platforms and Audio Only

During the COVID-19 Public Health Emergency (PHE), the DHHS Office exercised enforcement discretion to allow good faith use of telehealth. In doing so, DHHS relaxed the restriction of HIPAA compliant telehealth platforms and allowed the use of audio only (hhs.gov). As of March 30, 2023, platform requirements are re-established, and the provider must use a HIPAA compliant platform. Vendors that offer HIPAA compliant platforms and agreeable to a BAA include Skype for Business, Microsoft Teams, Zoom for Healthcare, and CISCO Webex Meetings and Webex Teams. The provider or practice owner should review workflows, update policies, and educate providers.

Other flexibilities have been made permanent. Behavioral/mental telehealth services can permanently be delivered using audio-only communication platforms to Medicare patients. DHHS OCR issued Guidance on How the HIPAA Rules Permit Covered Health Care Providers and Health Plans to Use Remote Communication Technologies for Audio-Only Telehealth (hhs.gov). The HIPAA security rules do not apply to traditional landlines, electronic communication technologies such as Voice over Internet Protocol (VoIP), and mobile technologies that use electronic media, such as the Internet, intra- and extranets, cellular, and Wi-Fi. Where the telecommunication provider only has transient access to ePHI, merely transmits ePHI, and functions as a conduit only, the provider is not required to enter into a business associate agreement. If there is greater than transient access to ePHI such as with session recording and transcripts in cloud-based storage, ePHI is being is received, stored, and maintained, so the provider would be required to enter a BAA to ensure compliance with HIPAA security rules. For similar reasons, a BAA would also be required with the developer of a smartphone app that the provider uses to translate oral communications to provide meaningful access to Limited English Proficiency Persons.

Audio-only has greater risks of impermissible uses or disclosure of PHI, particularly with identity verification and privacy of the setting. Verification of identity may be done orally or in writing, which may include electronic methods, or using automated real-time verification applications. If using an automated identity verification system, a BAA is required. Healthcare providers are expected to provide telehealth services in private settings. The provider must take reasonable precautions in office settings that include other healthcare professionals, such as shared offices. Also, as part of the consent process, the importance of the private setting to avoid inadvertent disclosures of PHI must be discussed with the patient. For adolescent patients in which smartphones and social media use is highly normalized, the provider should take care to emphasize the importance of the private setting to ensure no concurrent or future disclosure of PHI.

### 9.4.3 Tightening of the HIPAA Security Rules

On January 6, 2025, DHHS issued notice of proposed rulemaking (NPPRM) which would modify the Security Standards for the Protection of Electronic Protected Health Information ("Security Rule") under HIPAA and the HITECH Act (90 FR 898). The purpose of the proposed rule is to better protect the confidentiality, integrity, and availability of electronic protected health information (ePHI). If finalized, the new rule would have implications for telehealth providers. In the past, business associates were allowed to adopt security measures to fit their size, complexity, capabilities, and risk of breach (45 CFR § 164.306(b). This sliding scale, as it is commonly referred to, would no longer apply, and the business associate would be required to make annual certified reviews of their policies and procedures. Encryption of ePHI and multifactor authentication (MFA) with few exceptions, such as emergencies, would become mandatory. Security risk assessments would be required annually. Covered entities would also be required to establish and implement written incident response for occurrences that cause damage to systems containing ePHI, such as ransomware attacks. The new rule replaces the past technologically flexible approach with stringent security requirements. While having the potential to strengthen cybersecurity, these requirements will result in increased compliance costs for the telehealth provider.

## 9.5 Reimbursement Issues for the Use of Telehealth

The reimbursement system for telehealth consists of a labyrinth of requirements from federal laws, Medicare rules, varying state Medicaid statutes and regulations, and commercial payor contracts. Prior to the COVID-19 PHE, Medicare payment for telehealth services was restricted to rural locations, did not allow the originating site to include the home, contained a very narrow definition of eligible practitioners, and had to be provided by synchronous video, with audio-only not permitted except in rare circumstances (Shaver 2022). During the PHE, the CARES Act, CMS 1135 Waiver, and American Relief Act Congress established waivers and flexible enforcement policies to expand access to healthcare while containing the spread of the pandemic (Shaver 2022). These flexibilities were extended until March 30, 2025, through CMS rulemaking authority and Congressional continuing resolutions. The Full -Year Continuing Appropriations and Extensions Act, Public Law 119-4, passed on March 15, 2025, effectuated some permanent rule changes pertaining to behavioral telehealth. Audio-only delivery of behavioral health is permanent. Relaxation of geographic restrictions for behavioral health is also permanent, allowing the provider to deliver services to patients in another state subject to the state's licensure restrictions. Patients can receive behavioral/mental health services at home. The expansions for eligible providers to include mental health counselors, marriage counselors, and family therapists are also permanent. Other flexibilities survive by a fourth temporary extension (Consolidated Appropriations Act 2026) and are set to expire on December 31, 2027. (see Table 9.1, CMS Telehealth Regulations, Before and After the Public Health Emergency).

**Table 9.1** Summary of Center for Medicare and Medicaid Services (CMMS) Telehealth Regulations during and after the Public Health Emergency

| Before March 2020 | After CARES Act and CMS 1135 Waiver, American Relief Act (Ext. to 3/30/2025) | Legislative Extensions and Consolidated Appropriations Act, 2026 (H.R. 7148) |
|---|---|---|
| Who can perform and receive telehealth | | |
| Certain licensed providers | Any type of clinician who is eligible to bill CMMS (includes PT, OT, SLP, AUD, mental health counselors, marriage and family therapists) | Any type of clinician who is eligible to bill CMMS through 12/31/2027. Mental health counselors, marriage and family therapists **Permanent** |
| Patients and providers with a preexisting relationship (in-person visit requirement) | An in-person visit within 6 months of an *initial behavioral telehealth service*, and annually thereafter is not required | No in-person requirement for behavioral telehealth through 12/31/2027<br>For FQHCs and RHCs, the one in-person visit 6 months before behavioral/mental health and at least every 12 months will begin after 1/1/2028 |
| Location of telehealth | | |
| Specified sites only (designated rural areas, certain medical facilities) | Originate and conduct from any location in the United States and territories (no geographic restrictions) | No geographic restrictions (any location in the United States and territories) through 12/31/2027 |
| Physicians must conduct telehealth from their practice location (distant site) | Physicians may conduct telehealth from their home location | Physicians may conduct telehealth from their home location using their currently enrolled practice location for billing through 12/31/2027 |
| Patient must be at practitioner's office or specified medical facility, not at home (originating site) | Patients can get telehealth wherever they are located and at home | Patients can get nonbehavioral/mental health services at home until 12/31/2027<br>Patients can receive *behavioral/mental health* services at home **Permanent** |
| Telehealth may not cross state lines (patient's originating site) | No geographic restriction: telehealth can be provided to a patient in another state (except for state mandated restrictions) | No geographic restriction for nonbehavioral/mental health services through 12/31/2027 and for behavioral/mental health services, **permanent** |

(continued)

**Table 9.1** (continued)

| Before March 2020 | After CARES Act and CMS 1135 Waiver, American Relief Act (Ext. to 3/30/2025) | Legislative Extensions and Consolidated Appropriations Act, 2026 (H.R. 7148) |
|---|---|---|
| Type of technology | | |
| Audio-visual/video technology | *Behavioral health* can be delivered using audio only | *Behavioral health* can be delivered using audio only **Permanent** |
| Only approved technology platforms | Expanded approved platforms FaceTime, Skype, Zoom | |
| Reimbursement | | |
| Medicare coinsurance and deductibles apply | Providers may waive cost-sharing for telehealth paid for by government payors | First dollar deductible and cost sharing waivers expired |
| Lower reimbursement than for in-person services | Parity of reimbursement with in-person services, including audio only | Reimbursement parity extended through 12/31/2027 |
| "Direct supervision" requires presence on site (teaching physicians and incident-to billing) | "Direct supervision" includes virtual presence through real-time audio and visual interactive communications for certain services | Virtual "direct supervision" **Permanent** |

Adapted from Shaver (2022), Cares Act AMA Covid, https://www.ama-assn.org/health-care-advocacy/federal-advocacy/cares-act-ama-covid-19-pandemic-telehealth-fact-sheet; Medicare Telemedicine Healthcare Provider Fact Sheet, https://www.cms.gov/newsroom/fact-sheets/medicare-telemedicine-health-care-provider-fact-sheet; Telehealth Policy Updates, https://telehealth.hhs.gov/providers/telehealth-policy/telehealth-policy-updates; Consolidated Appropriations Act, 2026 (H.R. 7148); 42 CFR 410.26

Medicaid is funded by the state and federal government. States have flexibility to administer their Medicaid programs. As a result, Medicaid laws and regulations, including those for reimbursement of telehealth services, vary by state. Also, Medicaid regulation can be broader than Medicare. States have discretion on services to cover, types of providers, and restrictions on where the patient could be located and the way the remote services must be provided. Enrolled clinicians must refer to the state rules for the states (NCSL 2024).

The uncertain state of federal regulation can be a disincentive for practitioners to expand their investments in telehealth and related auxiliary modalities, increasing access barriers for patients. However, professional associations continue to advocate for the removal of telehealth barriers.

## 9.6 Telehealth Fraud and Abuse

The Department of Health and Human Services and the Office of the Inspector General (OIG) have substantially increased the scrutiny of telemedicine due to misuse companies and providers. As stated by Christi Grimm, HHS Principal Deputy

Inspector General, "OIG believes there is a shared goal: ensuring that telehealth delivers quality, convenient care for patients and is not compromised by fraud" (OIG Feb. 26, 2021). The OIG and the Department of Justice (DOJ); Criminal Fraud Division, which enforces the Fraud and Abuse laws; False Claims Act (FCA); and Anti-Kickback Statute (AKS) detect fraud in a variety of ways. Data analytics of Medicare and Medicaid billing data, hotline tips and whistleblowers all provide easily accessible information to high volume providers and potential abuse.

On July 20, 2022, the OIG issued a Special Fraud Alert regarding practitioners entering arrangements with telemedicine companies. The Special Fraud Alert described the indicia of problematic arrangements with practitioners and telemedicine companies. The companies are generally single purposes providers providing drugs or durable medical equipment. Patients are recruited through aggressive telemarketing, health fairs, TV, or social media advertising targeting Medicare beneficiaries. Often free items or services are offered. The practitioner is compensated on volume of items or services provided. The practitioner is required to follow practices that include minimum or no contact with the patient, impairing the provider's ability to meaningfully assess the medical necessity of the items or services ordered or prescribed. For example, practices include using exclusively audio-only technology regardless of patient preference and providing minimal medical history, usually limited to patient demographics and insufficient clinical information in a templated format. Visits are scripted with no opportunity for diagnostic testing, and there may be stringent time restrictions on visits. If the practitioner takes more time, they no longer receive referrals. Since the company furnishes only one product or a single class of products (such as DME or a specific drug), the course of treatment is predetermined, and the practitioner has few options. As such, the practitioner may be pressured to practice outside their scope or comfort zone either through prescribing practices or licensure in other states. Typically, there is no follow-up with the patients (OIG July 20, 2022). The conditions place the practitioner in the position of inadequate documentation for a reimbursable telemedicine service and potential professional liability due to the inadequate history, abbreviated options, and lack of continuity of care.

DOJ pursued millions of dollars of healthcare fraud associated with telehealth in 2023 and 2024. The DOJ indicted Ruthia He, founder and CEO of Done Global Inc. for $100 million in Adderall Distribution and healthcare fraud scheme using telemedicine (Department of Justice Press Release June 13, 2024a). David Brody, Clinical President, was indicted, also. The case was the DOJ's first criminal drug prosecution related to a digital health company that distributed controlled substance via telemedicine. On the surface Done Global was a large, seemingly reputable company. It generated over $100 million in revenue arranging for the prescription of over 40 million Adderall and other stimulants and Scheduled II controlled substances. The DOJ alleged that DONE targeted drug seekers through deceptive advertising and provided easy access to Adderall for a monthly prescription fee. The company allegedly limited information available to prescribers and instructed them to prescribe Adderall even if the member did not qualify. DONE required encounters to be under 30 minutes and paid the providers lucratively. The compensation structure discouraged follow-up care, and an auto-refill function, not permissible for Schedule II controlled drugs, was used. At least one death is associated with overdosing due to prescribing practices. Ruthia HE and David Brody were charged with conspiracy to distribute controlled substances and drug trafficking of controlled substances without a legitimate medical purpose and outside the usual course of professional practice. Fraud charges were also brought against physicians, nurse practitioners, and other licensed healthcare professionals who worked for DONE Global.

DONE's primary competitor in the telehealth market, Cerebral, Inc., was considered by many to be a reputable online mental health company. Similar to DONE Global, its business model relied upon aggressive marketing practices and featured dancing pill bottles and a celebrity athlete. Cerebral correlated undiagnosed ADHD and young women who were binge eating or overeating. Targeting populations seeking weight loss, Cerebral created policies to increase prescriptions of ADHD medications, such as Adderall. The company disciplined low prescribing physicians and paid bonuses to high prescribers. Under pressure of indictment by the DOJ, Cerebral entered into a non-prosecution agreement and agreed to pay over $3.6 million, the amount attributed to monthly increase in ADHD revenue generated from the prescribing. In entering the Non-Prosecution Agreement, the DOJ recognized that the company agreed to mitigate its practices by no longer prescribing controlled substances to is patients, cooperated with in the investigation with detailed disclosures and had no prior criminal record (Department of Justice Press Release Nov 4, 2024b).

Another fraud scheme, known as Doctor Chase, involves the theft of the provider's NPI. The provider is sent faxes that look like prescription refills with the expectation that staff will sign electronically on behalf of the provider and return them. The scripts are then sold and resold before being filled at a pharmacy that does the billing. The provider's front and back office staff should be trained to be alert to and report the receipt of unrecognizable faxes or other electronic communications asking for the provider's signature (Langley et al. 2025).

Professional liability insurance policies either entirely exclude allegations of fraud or provide a defense only under a reservation of rights for criminal fraud cases. The provider would be obligated to repay defense costs if found guilty and pay any settlement, judgment, penalty, or fine. It is against public policy to make an agreement of indemnity to cover settlements or court judgments in a criminal fraud case. Any such payments, such as a non-prosecution agreement, are out of pocket costs for the provider or company. Risk management necessitates stringent policies for due diligence when providers seek to contract with a telemedicine company, security of provider billing information such as NPIs, and education of providers and support staff.

## 9.7 Conclusion

Telehealth presents a viable alternative to in-person care and is increasingly used for the ongoing treatment of ADHD. Improved access, convenience, especially for busy families of children with ADHD, and avoidance of stigma encourage the expansions of telehealth use. The clinician using telehealth must be fastidious about the professional requirements, broad regulation, and business considerations as they embark on telehealth practice. Professional advocacy is needed to help streamline and standardize policy affecting telehealth, with the goal to ensure access for populations requiring ongoing treatment and management.

**Acknowledgment** This chapter benefitted from the unfaltering and responsive efforts of Margaret Murolo JD, RN, who expertly researched a conglomeration of laws and regulations and eagerly met all of my requests.

## Appendices

### Appendix 9.1: Professional Requirements for Telehealth

- Type of health care professional authorized to provide telehealth services.
- Licensed in the state where the patient is located (originating state) and where practitioner is located (distant site) at the time of service.
- Follow procedures of Interstate Compact or streamlined state registrations, if applicable.
- If prescribing controlled substances, obtain federal DEA registration in each state where the patient is located.
- Obtain DEA Telemedicine Prescribing Authorization Registration or Advanced Telemedicine Prescribing Authorization Registration in each state, if applicable.
- Same standard of care as in-person setting.
- Scope of practice: Complies with laws and regulations of the state's licensing board.
- Informed Consent from patient for use of telehealth; use teach-back for understanding risks and contingency plans.
- Provide reasonable accommodation or auxiliary aids for special populations such as Limited English Proficient Persons, deaf, cognitively or vision impaired persons, or other disability impacting telehealth accessibility.
- Use HIPAA-compliant platform and obtain BAA from all vendors, including software, remote patient monitoring (RPM) or auxiliary devices, and analytics.
- Confirm or obtain professional liability insurance telehealth coverages.

## Appendix 9.2: Telehealth Consent Teach-Back Documentation (Example)

Name of patient:__________________________________________ Date:_________

Name of clinician:_________________________________________

Name of interpreter:________________________________________

Use of auxiliary aid (specify):______________________________

(If patient is Limited English Proficient Person (LEPP), has deafness, vision impair ment, or other disability, specify the reasonable accommodation or auxiliary aid provided.)

Application(s)____________________________________________

Was the patient able to teachback in their own words?

| | Attempt 1 | Attempt 2 | Attempt 3 |
|---|---|---|---|
| What telehealth is | Yes_ No__ | Yes_ No__ | Yes_ No__ |
| How telehealth could help (benefits) | Yes_ No__ | Yes_ No__ | Yes_ No__ |
| How telehealth could be bad (harm/risks) | Yes_ No__ | Yes_ No__ | Yes_ No__ |
| Privacy of telehealth (risks) | Yes_ No__ | Yes_ No__ | Yes_ No__ |
| Alternative of office visit | Yes_ No__ | Yes_ No__ | Yes_ No__ |
| Cost | Yes_ No__ | Yes_ No__ | Yes_ No__ |
| How to use application | Yes_ No__ | Yes_ No__ | Yes_ No__ |
| Understands plan if technology fails | Yes_ No__ | Yes_ No__ | Yes_ No__ |
| Emergency or safety care plan | Yes_ No__ | Yes_ No__ | Yes_ No__ |
| Ability to withdraw consent | Yes_ No__ | Yes_ No__ | Yes_ No__ |

Sign if decide to use telehealth (voluntary)___________________________

Adapted from AHRQ, 2020, https://www.ahrq.gov/health-literacy/improve/informed-consent/index.html; and 45 C.F.R § 92.104(a), American Psychological Association Guidelines for the Practice of Telehealth (2024). **The consent form should be tailored to specific state requirements.**

## References

Affordable Care Act, Section 1557, codified at 42 U.S.C. § 18116. Available at: https://codes.findlaw.com/us/title-42-the-public-health-and-welfare/42-usc-sect-18116/.

AHRQ's Easy-To-Understand Telehealth Consent Form, Content last reviewed September 2020. Agency for healthcare research and quality, Rockville, MD. Available at: https://www.ahrq.gov/health-literacy/improve/informed-consent/index.html.

American Psychological Association. Proposed revision of guidelines for the practice of telepsychology. 2024. Available at: https://www.apa.org/practice/guidelines/telepsychology-revision.pdf.

American Relief Act. H.R.10545 (118th) Public Law 118-158. 2025. Available at: https://www.govtrack.us/congress/bills/118/hr10545.

American Telemedicine Association, Telehealth Basics. Available at: https://www.americantelemed.org/resource/why-telemedicine/. Accessed 1 June 2025.

Brokamp v. James, 66 F.4th 374 (2d Cir. 2023).

Center for Connected Health Policy (CCHP). Reports: state telehealth laws and reimbursement policies report. Fall 2024. Available at: https://www.cchpca.org/resources/state-telehealth-laws-and-reimbursement-policies-report-fall-2024/. Accessed 1 June 2025.

Center for Connected Health Policy (CCHP). Fall 2025 Report: state telehealth laws and medicaid program policies, interactive policy tool. 2025. Available at: https://www.cchpca.org/all-telehealth-policies/. Accessed 30 Dec 2025.

Center for Connected Health Policy (CCHP), Consent, Professional Requirements. Available at: https://www.cchpca.org/topic/consent-requirements-professional-requirements/. Accessed 1 June 2025.

Center for Connected Health Policy (CCHP), Definitions, Professional Requirements. Available at: https://www.cchpca.org/topic/definitions-professional-requirements/.

Center for Connected Health Policy (CCHP) On-line prescribing, Professional Requirements. Available at: https://www.cchpca.org/topic/online-prescribing/. Accessed 1 June 2025.

Center for Connected Policy (CCHP), Licensure Compacts. Available at: https://www.cchpca.org/topic/licensure-compacts/. Accessed 1 June 2025.

Center for Medicare and Medicaid Services (CMS), Section1135, Coronavirus Waivers and Flexibilities. Available at: https://www.cms.gov/coronavirus-waivers. Accessed 1 June 2025.

CNO/NSO. Nurse practitioner spotlight: telemedicine. 2023. Available at: https://www.nso.com/getmedia/f4aabaa6-36c0-4421-8061-2ca98bf0b91f/NSO_NP23_SL_Telemedicine_SEC-(1).pdf.

Comphealth. Credentialing and licensing: interstate medical licensure compact states list and guide for 2025. 2025. Available at: https://comphealth.com/resources/interstate-medical-licensure-compact.

Confidentiality of Substance Abuse Disorder Patient Records, 42 U.S.C.§290dd-2.

Coronavirus Aid, Relief and Economic Security (CARES) Act, (116th Congress) P.L.116-136 (codified at 15 U.S.C. Chapter116).

Danielson ML, Claussen AH, Bitsko RH, Katz SM, Newsome K, Blumberg SJ, Kogan MD, Ghandour R. ADHD prevalence among U.S. children and adolescents in 2022: diagnosis, severity, co-occurring disorders, and treatment. J Clin Child Adolesc Psychol. 2024;53(3):343–60. https://doi.org/10.1080/15374416.2024.2335625.

Department of Justice. Press release, justice department's first criminal drug distribution prosecutions related to a digital health company that distributed controlled substances via telemedicine. June 13, 2024a. Available at https://www.justice.gov/usao-ndca/pr/founderceo-and-clinical-president-digital-health-company-arrested-100m-adderall.

Department of Justice. Press release, telehealth company cerebral agrees to pay over $3.6 million in connection with business practices that encouraged the unauthorized distribution of controlled substances. November 4, 2024b. Available at: https://www.justice.gov/usao-edny/pr/telehealth-company-cerebral-agrees-pay-over-36-million-connection-business-practices.

Drug Enforcement Agency and Department of Health and Human Services. Separate Registrations for Separate Locations, 90 Fed. Reg. 6541, 6550 (codified to 21 C.F.R. §1301.12). January 17, 2025. Available at https://www.ecfr.gov/current/title-21/chapter-II/part-1301/subject-group-ECFR0f5a129834f0129/section-1301.12.

Elizabeth Brokamp v. Leticia James, in her official capacity as Attorney General of New York, (U.S.C.A. 2nd Circuit), No. 21-3050 (April 27, 2023), cert. denied U.S. March 2024. Available at: https://law.justia.com/cases/federal/appellate-courts/ca2/21-3050/21-3050-2023-04-27.html.

Federal Trade Commission. Health breach notification rule, 89 Fed. Reg.47028 (codified at 16 C.F.R Part 318). May 30, 2024. Available at: https://www.federalregister.gov/documents/2024/05/30/2024-10855/health-breach-notification-rule. Accessed 1 June 2025.

Fogel AL, Kvedar JC. Reported cases of medical malpractice in direct-to consumer -telemedicine. JAMA. 2019;321(13):1309–10. Available at: https://jamanetwork.com/journals/jama/fullarticle/2729359.

Full-Year Continuing Appropriations and Extension Act, 119th Congress. Public Law No. 119-4. March 15, 2025. Retrieved from https://www.congress.gov/119/plaws/publ4/PLAW-119publ4.pdf.

Galvin E, Gavin B, Kilbride K, Desselle S, McNicholas F, Cullinan S, Hayden J. The use of telehealth in attention-deficit/hyperactivity disorder: a survey of parents and caregivers. Eur Child Adolesc Psychiatry. 2024;33:4247–57. https://doi.org/10.1007/s00787-024-02466-y. National Library of Medicine. Available at https://pubmed.ncbi.nlm.nih.gov/38753037/.

Giovannini S, Watkins S. That's so metadata, navigating the nuances of privacy compliance, cybersecurity, and sensitive data collection for telehealth businesses and healthcare organizations, telemedicine: legal and compliance issues conference. Washington, D.C.: American Health Law Association; 2025.

Hawaii Revised Statues. §453-1.3 Practice of telehealth. 2024. Retrieved from https://law.justia.com/codes/hawaii/title-25/chapter-453/section-453-1-3/.

Health Information Technology for Economic and Clinical Health (HITECH) Act of 2 of 2009, Public Law 111-5 (codified at 42 USC sec 139w-4(0)(2) and 42 USC sec 139w-4(0)(2)).

Health Insurance Portability and Accountability Act of 1996a (HIPAA), Public Law 104-191, (codified at Title 42 of the U.S.C., §1320d through 1320d-8).

Health Insurance Portability and Accountability Act of 1996b (HIPAA), Key Definitions, 45 C.F.R.§160.103.

Hyder MA, Razzak J. Telemedicine in the United States: an introduction for students and residents. J Med Internet Res. 2020;22(11):e20839. https://doi.org/10.2196/20839.

Institute of Medicine. Crossing the quality chasm: a new health system for the 21st century. Natl Libr Med. 2001; https://doi.org/10.17226/10027. Available at https://pubmed.ncbi.nlm.nih.gov/25057539/.

Langarizadeh M, Tabatabaei MS, Tavakol K, Naghipour M, Rostam A, Moghbel F. Telemental health care, an effective alternative to conventional mental care: a systematic review. Acta Inform Med. 2017;25(4):240–6. https://doi.org/10.5455/aim.2017.25.240-246.

Langley T, O'Neal DA, Wagner C. The promise and peril of telehealth: practice, reimbursement and government enforcement, legal and compliance issues virtual conference. Washington D.C.: American Health Law Association; 2025.

Legislative Extensions and Consolidated Appropriations Act, 2026, H.R. 7148, 119th Cong. (2026).

Maheu, M. (2024). Controversy: mandating the California PSYPACT through legislation?, Policy and Compliance, telehealth.org. Available at https://telehealth.org/blog/implement-the-california-psypact/..

Moore PJ. Medical malpractice: the effect of doctor-patient relations on medical patient perceptions and malpractice intentions. West J Med. 2000;173:224–50. National Library of Medicine. Available at: https://pubmed.ncbi.nlm.nih.gov/11017984/.

National Council of State Boards of Nursing (NCSBN), Nurse Licensure Compact. Available at: https://www.nursecompact.com.

National Council of State Boards of Nursing (NCSBN), APRN Licensure Compact. Available at: https://www.ncsbn.org/nursing-regulation/practice/aprn.page.

National Council of State Legislatures, Telehealth Explainer Series: Licensure Flexibilities for Telehealth (Oct. 24, 2024). from https://www.ncsl.org/health/the-telehealth-explainer-series/licensure-and-interstate-compacts. Accessed 28 May 2025.

Northeast Telehealth Resource Center. HIPAA and telehealth, a stepwise guide to compliance. April 24 2015. Retrieved from https://netrc.org/wp-content/uploads/2015/11/HIPAA-Stepwise-Guide.pdf.

Office of Civil Rights and Department of Justice Civil Rights Division. Guidance on nondiscrimination in telehealth, Department of Health and Human Services. April 11, 2025. Available at https://www.hhs.gov/civil-rights/for-individuals/disability/guidance-on-nondiscrimination-in-telehealth/index.html. Accessed 1 June 2025.

Office of the Inspector General (OIG). Media release: statement of HHS principal deputy inspector general, Christi Grimm, Department of Health and Human Services. February 26, 2021. Available at https://oig.hhs.gov/coronavirus/letter-grimm-02262021.asp. Accessed 1 June 2025.

Office of the Inspector General (OIG). Special fraud alert: OIG alerts practitioners to exercise caution when entering arrangements with telemedicine companies. Department of Health and Human Services Special. July 20, 2022. Available at https://oig.hhs.gov/documents/root/1045/sfa-telefraud.pdf. Accessed 1 June 2025.

Ortiz N, Preuss CV. The controlled substances act, continuing education activity. National Library of Medicine; 2024. https://www.ncbi.nlm.nih.gov/books/NBK574544/.

Patton K, Macon C, Jain S. Telehealth demand: an update four years after the onset of the COVID-19 pandemic. The Compass, Trilliant Health; 2024. Available at https://www.trillianthealth.com/market-research/studies/telehealth-demand-an-update-four-years-after-the-onset-of-the-covid-19-pandemic.

PSYPACTMap® Participating States. https://psypact.gov/page/psypactmap.

Rowland SP, Fitgerald JE, Lungren M, Lee EH, Harned Z, McGregor AH. Digital health technology-specific risks for medical malpractice liability. NPJ Digit Med. 2022;5:157. https://doi.org/10.1038/s41746-022-00698-3. Accessed 1 June 2025.

Sean McBride M.D. and Shellye Horowitz, M.D. v. Randy Hawkins, in his official capacity as President of the Medical Board of California. (E.D. Calif), No. 2:24-CV-01394. May 16, 2024. Available at https://cases.justia.com/federal/district-courts/california/caedce/2:2024cv01394/446346/20/0.pdf. Accessed 28 May 2025.

Section 504 of the Rehabilitation Act of 1973, codified at 29 U.S.C.§794. Retrieved from https://codes.findlaw.com/us/title-29-labor/29-usc-sect-794/.

Shah P, Thornton I, Kopitnik NL, Hipskind JE. Informed consent, continuing education activity. National Library of Medicine; 2024. Retrieved from https://www.ncbi.nlm.nih.gov/books/NBK430827/.

Shannon McDonald M.D., Paul Gardner, M.D.; J.A., a minor, by and through guardian and next friend Michael Abell; Michael Abell; and Hank Jennings v. Otto F. SABANDO, in his official capacity as President of the New Jersey State Board of Medical Examiners, Defendant, (USDC, D. New Jersey) No. 3:23-CV-23044 (December 13, 2023). 2023 WL 11780536.

Shaver JM. The state of telehealth before and after the COVID-19 pandemic. Prim Care Clinic Office Pract. 2022;49:517–30. https://doi.org/10.1016/j.pop.2022.04.002.

Spencer T, Noyes E, Biederman J. Telemedicine in the management of ADHD: literature review of telemedicine in ADHD. J Atten Disord. 2019;24(1) SAGE Jounrals https://doi.org/10.1177/1087054719859081.

Staley BS, Robinson LR, Claussen AH, Katz SM, Danielson ML, Summers AD, Farr SL, Blumberg SJ, Tinker SC. Attention deficit hyperactivity disorder, diagnosis, treatment and telehealth use in adults. MMWR. 2023;73(40):890–5. Available at https://www.cdc.gov/mmwr/volumes/73/wr/mm7340a1.htm.

Title 21, United States Code, § 829e, Prescriptions, Controlled substances dispensed by means of the Internet. Available at https://uscode.house.gov/view.xhtml?req=granuleid:USC-prelim-title21-.

Title 21 Code of Federal Regulations, §1307.4.(Temporary extension of certain COVID-19 telemedicine flexibilities for prescription of controlled medications. Available at https://

www.ecfr.gov/current/title-21/chapter-II/part-1307/subject-group-ECFR7bd0454792d3c77/section-1307.41#.

Title 45 Code of Federal Regulations FR § 164.514, Other requirements relating to uses and disclosures of protected health information. Available at https://www.ecfr.gov/current/title-45/subtitle-A/subchapter-C/part-164/subpart-E/section-164.514.

U.S. Department of Health and Human Services (DHHS). Notice of proposed rulemaking: HIPAA rule to strengthen the cybersecurity of electronic protected health information, 90 Fed. Reg. 898 (Jan. 6, 2025). (Amending 45 C.F.R §160 and 164). Available at https://www.federalregister.gov/documents/2025/01/06/2024-30983/hipaa-security-rule-to-strengthen-the-cybersecurity-of-electronic-protected-health-information.

U.S. Department of Health and Human Services (DHHS). Guidance on How the HIPAA Rules Permit Covered Health Care Providers and Health Plans to Use Remote Communication Technologies for Audio-Only Telehealth. Available at https://www.hhs.gov/hipaa/for-professionals/privacy/guidance/hipaa-audio-telehealth/index.html.

U.S. Department of Health and Human Services (DHHS), Office of Civil Rights (DHHS-OCR). Guidance on Discrimination in Telehealth: Federal Protections to Ensure Accessibility to People with Disabilities and Limited English Proficient Persons. 2025. Available at https://www.hhs.gov/civil-rights/for-individuals/disability/guidance-on-nondiscrimination-in-telehealth/index.html

U.S. Department of Health and Human Services, Office of Inspector General. Special Fraud Alert: OIG Alerts Practitioners to Exercise Caution When Entering into Arrangements with Purported Telemedicine Companies. July 20, 2022. Available at https://oig.hhs.gov/documents/root/1045/sfa-telefraud.pdf.

U.S. Department of Justice and Drug Enforcement Administration. Third Temporary Extension of COVID-19 Telemedicine Flexibilities for Prescription of Controlled Medications, 89 Fed. Reg. 91253 (to be codified at 21 C.F.R. 1307, 42 C.F.R. Part 12). November 19, 2024. Available at https://www.federalregister.gov/documents/2024/11/19/2024-27018/third-temporary-extension-of-covid-19-telemedicine-flexibilities-for-prescription-of-controlled#p-amd-2.

U.S. Drug Enforcement Administration (DEA). Notice of Proposed Rulemaking: Special Registrations for Telemedicine and Limited State Registrations, 90 Fed. Reg.6541 (Jan. 17, 2025). (Amending 21 C.F.R §1300, 1301, 1304 and 1306). January 17, 2025. Available at https://www.federalregister.gov/documents/2025/01/17/2025-01099/special-registrations-for-telemedicine-and-limited-state-telemedicine-registrations.

Uniform Law Commission. Uniform Telehealth Act, 2022a. Available at https://www.uniformlaws.org/viewdocument/enactment-kit. Accessed 29 June 2025.

Uniform Law Commission. Uniform Telehealth Act Summary. 2022b. Available at https://www.uniformlaws.org/search?executeSearch=true&SearchTerm=ULC+Telehealth+Act+Summary&l=1.

Uniform Law Commission (also known as National Conference of Commissioners on Uniform State Laws). Available at https://www.uniformlaws.org/home. Accessed 29 June 2025.

Uniform Law Commission, Telehealth Act, Enactment Map. Available at https://www.uniformlaws.org/committees/community-home?CommunityKey=2348c20a-b645-4302-aa5d-9ebf239055bf.

Venditti SA, Sazegar P, Fuchs LC, Snarskis CE. Family medicine resident and faculty perceptions about strengths and limitations of telemedicine. PriMER. 2022;2022:6–9. National Library of Medicine. Available at https://doi.org/10.22454/primer.2022.665996. Accessed 1 June 2025.

Wolf TD. Telemedicine and malpractice: creating uniformity at the national level. William & Mary Law Review. 2020;61(5):1504–36. Available at https://scholarship.law.wm.edu/wmlr/vol61/iss5/7/.

# Index



© The Editor(s) (if applicable) and The Author(s), under exclusive license to Springer Nature Switzerland AG 2026
R. J. Kennedy (ed.), *Assessment & Treatment of ADHD*,
https://doi.org/10.1007/978-3-032-24171-9